TOBACCO

TOBACCO

A Cultural History of
How an Exotic Plant
Seduced Civilization

Iain Gately

GROVE PRESS
New York

First published in Great Britain in 2001 by
Simon & Schuster UK Ltd., London, England

Printed in the United States of America

Library of Congress Cataloging-in-Publication Data

Gately, Iain, 1965–
    Tobacco : a cultural history of how an exotic plant seduced civilization /
Iain Gately.
        p. cm.
    First published in London by Simon & Schuster, 2001.
    Includes bibliographical references (p.).
    ISBN 0-8021-3960-4 (pbk.)
    1. Tobacco—History. 2. Tobacco—Social aspects—History. I. Title.

SB273 .G29 2002
394.1'4—dc21   2001054493

Grove Press
841 Broadway
New York, NY 10003

04 05 06 07   10 9 8 7 6 5 4 3

Acknowledgements

I would like to thank Matthew Taylor for his intelligent and diligent research, in particular his astute assessments of the present state of the tobacco industry, Rocio Olid Fiances for providing me with material relating to tobacco use in Spain, James Johnstone for the title and Fiona Gately for My Lady Nicotine. *Hecho en Tarifa.*

Contents

The Mayan god of the underworld enjoys a cigar at
Xac Pal's coronation as Sky Penis.

1

To Breathe is to Inhale

*The discovery of tobacco and smoking – the
social, ritual and medical roles of tobacco in
South, Central and North America – tobacco
pipes and their official functions*

Why smoke? I drifted into the habit in the year between my
eighteenth and nineteenth birthdays. I lived in Hong Kong
at the time where smoking was so commonplace as to appear
a natural act, not a habit. The first time I thought about it
at all was on a train through south-east China, travelling in
a third class carriage with a non-smoking friend. A middle-
aged Chinese offered us both cigarettes – eager to break the
ice and to practise his English. My travelling companion,
Philip, declined, which amazed the Chinese.

'Not want now?' he inquired, with a look of concern for
Philip's sanity.

'No thanks. I don't smoke.'

'Your friend not want cigarette.'

'He never smokes.'

'Mean: not smoke now?' The Chinese was becoming
defensive. Communicating with Westerners was more
complex than he'd imagined.

'Not smoke ever.'

'Not smoke on train?'

'Not smoke.'

'Not smoke yet today? Here! Have first one!'

Meeting someone who could not conceive that people existed who did not smoke made me curious as to why tobacco has such a hold on mankind – why people who have never seen a cigarette advert or watched a Grand Prix and whose lifestyle choices are limited to survival imperatives prefer tobacco to food – prefer stimulation ahead of nourishment.

Why has smoking been so readily accepted into so many different cultures, where it has been the subject of creation myths and demonologies? What is the secret of its strange compulsion, which causes experiment to lead to slavery? And why, ultimately, a generation after the practice has been revealed as a killer, does it persist, and even multiply?

Tobacco and mankind have been associated since prehistory, and in a manner, while not unique, unlike most human–vegetable relationships. Tobacco is raised to be burned. It is bred to stimulate our lungs, not feed our stomachs. An investigation of the answers to the question 'Why smoke?' must begin by looking at a prior question: 'Why tobacco?' Why are we such an excellent match for each other? And why is smoking the usual way of celebrating our friendship? Mankind smokes other plants, including cannabis and opium poppies, but none with such frequency or such ubiquity as tobacco. Tea, for instance, which resembles tobacco in the sense that it is consumed for stimulation, not nourishment, is rarely smoked.

Although the tobacco plant is prettier than the tea bush, there is nothing in its appearance that singles it out for smoking. Tobacco's genus, *Nicotiana*, contains sixty-four species, two of which are involved in the affair with mankind, *Nicotiana rustica* and *Nicotiana tabacum*. Most tobacco presently consumed by humans is *Nicotiana tabacum*, a tall, annual, broad leafed plant. *Nicotiana rustica* is similar in appearance, but shorter and with fleshier leaves. They are attractive plants to look at – when confronted by a tobacco plant even botanists' prose becomes voluptuous. Here, for example, is *Nicotiana tabacum*'s debut in Gerard's *Herball* in 1636:

Tobacco, or henbane of Peru, hath very great stalks of
the bigness of a child's arm, growing in fertile and well
dunged ground seven or eight feet high, dividing itself
into sundry branches of great length, whereon are
placed in comely order very fair, long leaves, broad,
smooth and sharp pointed; soft and of a light green
colour; so fastened about the stalk that they seem to
embrace it. The flowers grow at the top of the stalks,
in shape like a bell flower, somewhat long and cornered,
hollow within, of a light carnation colour, tending to
whiteness towards the brim . . . the root is great, thick
and of a woody substance, with some thready strings
annexed thereto.

Both *Nicotiana rustica* and *Nicotiana tabacum* are native
only to the Americas, where mankind came across them
about 18,000 years ago. The humans who first populated
the American continent did not know tobacco and did not
smoke. They were of Asiatic origin and after crossing the
Bering Strait land bridge, they dispersed southwards through
the continent. In those areas where the fauna and terrain
most closely resembled that which they had left behind, they
continued a nomadic existence. To the south, however, they
cultivated vegetables, built cities, framed laws and gained
empires through conquest. Both nomads and settlers shared
an ancestral knowledge of herbs, which they augmented with
the new plants they encountered. Tobacco was one of these.
The discovery itself was unremarkable in a list of finds that
included such everyday consumables as potatoes, tomatoes,
rubber, chocolate and maize.

Plant geneticists have established that tobacco's 'centre of
origin', i.e. the meeting place between a species' genetic origin
and the area in which it was first cultivated, is located in the
Peruvian/Ecuadorean Andes. Estimates for its first date of
cultivation range from 5000–3000 BC. Tobacco use then
spread northwards and by the time of Christopher Columbus's
arrival in 1492 it had reached every corner of the American
continent, including offshore islands such as Cuba.

Quite how humans became interested in tobacco is unknown. Our ancestors were certainly open minded about diet and probably adopted an 'eat it then find out' approach. It is, however, certain that early Americans invented a new method of consumption for their herbal friend: smoking. That lungs had a dual function – could be used for stimulation in addition to respiration – is one of the American continent's most significant contributions to civilization. Human lungs have a giant area of absorbent tissue, every inch of which is serviced by at least a thousand thread-like blood vessels, which carry oxygen, poisons and inspiration from the heart to the brain. Their osmotic capacity is over fifty times that of the human palate or colon. Smoking is the quickest way into the blood stream short of a hypodermic needle.

Anthropologists have speculated as to how the discovery was made that smoke could be a source of pleasure as well as an irritant and have postulated several proto-smokers along Promethean lines. A typical *mise-en-scène* of this cultural landmark involves an ancestor striding through the ashes of a forest he has just burned down when he trips and falls headfirst into a smouldering bush of *Nicotiana rustica*. Although injured, he is soothed by the burning herb and adopts the habit of inhaling upon recovery.

It is more likely that the practice of smoking evolved from snuffing, i.e. inhaling powdered tobacco through the nose. Snuffing tubes are among the most ancient tobacco-related artefacts discovered in the Americas and the practice coexisted with smoking in South and Central America. Snuffing as a human habit was unique to the Americas, whose inhabitants seem to have considered the nose as a more versatile object than a bifurcated passage for air. They snuffed through their noses, smoked through their noses and even drank through their noses. It is tempting to imagine that the ever resourceful Americans, having conquered the nasal passages, perceived their lungs as the next challenge.

Smoking was only one of many tobacco habits in South

America. Beginning at tobacco's centre of origin around the Andes and tracing its progress north, the most striking features of early tobacco use are the variety of reasons employed to justify its consumption, and the diversity of ways in which it was taken. Tobacco was sniffed, chewed, eaten, drunk, smeared over bodies, used in eye drops and enemas, and smoked. It was blown into warriors' faces before battle, over fields before planting and over women prior to sex, it was offered to the gods, and accepted as their gift, and not least it served as a simple narcotic for daily use by men and women. Tobacco's popularity is in part explained by its biphasic nature as a drug. A small quantity of tobacco has a mild effect on its user, whereas in large doses it produces hallucinations, trances and sometimes death.

Many of the external applications of tobacco such as fumigation of crops and virgins were justified on practical grounds. Tobacco is a powerful insecticide, and blowing smoke over seed corn or fruit trees was an effective method of pest control. Some South American tribes also applied tobacco juice directly to their skin to kill lice and other parasites. These real qualities were embellished with mythical properties, so that tobacco came to be associated with cleansing and fertility, hence its application to maidens on their wedding night. As a result of its use in the planting season, tobacco became linked with initiation, and was adopted by many tribes as a symbol of the rites of passage between puberty and adulthood. For example, the Tucano of the north-west Amazon would give tobacco snuff to adolescent boys before they were 'presented formally to the sacred trumpets as newly initiated men'. It is fascinating to note that even in ancient civilizations tobacco was considered to be something that youth should aspire to use – it was part of being grown-up, and children yearned for the day when they would be treated as adults and be allowed to smoke.

Perhaps the most important use of tobacco in South American societies was as a medicine. Its mild analgesic and

antiseptic properties rendered it ideal for treatment of minor
ailments such as toothache, when its leaves would be packed
around the affected tooth, or wounds, when leaves or tobacco
juice would be applied to the area. It was further believed
to be an effective remedy for snake bites and, by extension,
a charm to ward off snakes. In addition to healing such
straightforward ailments, tobacco was employed to cure
serious illnesses, and to comprehend its perceived virtues as
a cure for fever, or cancer, it is necessary to examine the
South American Indian conception of disease. They believed
that diseases were caused by supernatural forces, in one of
two manners. These were either: (1) intrusion – a form of
possession, whereby an evil spirit or object had entered the
body of the sufferer, making them ill; or (2) soul loss, whereby
'the sufferer's soul was believed to be drawn away, and/or
to have wandered off into reaches of the supernatural world,
often into the land of the dead'. In order to be capable of
curing diseases defined in these terms, South American witch
doctors, or shamans, underwent a rigorous spiritual training
to enable them to undertake 'vision quests', in the course of
which they might identify the cause of the disease, and either
eject the evil intruder, or retrieve the wandering soul, and
thus restore the sufferer to health.

Tobacco played a central role in the spiritual training of
shamans. In the right doses, tobacco is a dangerously
powerful drug and a fatal poison. Shamans used tobacco,
often in conjunction with other narcotics, to achieve a state
of near death, in the belief that 'he who overcomes death
by healing himself is capable of curing and revitalizing
others'. Shamans undergoing initiation training were
required to take enough tobacco to bring them to the edge
of the grave.

The spiritual journeys undertaken by initiate shamans
were perceived as real quests, during the course of which
the neophyte would encounter terrible hazards. The priest
shaman of the Warao, for example, endured a series of perils
similar to those set out in computer games. After clearing
an abyss 'filled with hungry jaguars, snapping alligators, and

frenzied sharks, all eager to devour him' the tobacco intoxicated neophyte had to

> pass places where demons armed with spears are
> waiting to kill him, where slippery spots threaten to
> unbalance, and where giant raptors claw him. Finally,
> he must pass through a hole in an enormous tree with
> rapidly opening and closing doors. These symplegades
> are the actual threshold between life and death.
> Jumping through the clashing doors, he beholds the
> bones of those who went before him but failed to clear
> the gateway. Not finding his own bones among them
> he returns from the other-world restored to new life.

A tobacco shaman used the weed in almost every aspect of
his art. Tobacco smoke was employed as a diagnostic tool to
examine sick patients, and formed a part of many ceremonies
over which these doctor-priests officiated. Ritual smoke
blowing, by which a shaman might bestow a blessing or
protection against enemies both real and invisible, was
intended to symbolize a transformation, in which the tobacco
smoke represented a guiding spirit, and thus is reminiscent of
Christian ritual, whereby wine and bread are transubstantiated by a priest into the body and blood of Christ himself.
Shamans therefore were early proponents of passive smoking,
which they believed to be a force for good for non-smokers.

Turning to the methods by which tobacco was consumed
in South America, the astonishing diversity of tobacco habits
reflects not only the multitudinous purposes it served, but
also the different climatic conditions in which the weed was
employed. For instance, it was hard to smoke in the thin,
dry air of the Andes, so snuffing tended to prevail. Similarly,
in the swamplands of the Amazon, where fires could not be
kindled readily, tobacco was taken as a drink. Different
methods of tobacco consumption often existed side by side
– one form for everyday use, another for magic or ritual.

Probably the oldest way of taking the weed, and the most
straightforward, was chewing it. Cured tobacco leaves were

mixed with salt or ashes, formed into pellets or rolls, then tucked into the user's cheek, or under a lip. The juices thus released then dissolved in saliva and slid down the masticator's throat. Tobacco chewing could be recreational, or magical. The next method of consumption, in terms of complexity and pedigree, was drinking tobacco, in a sort of tea. Tobacco leaves were boiled or steeped in water and the resulting brew drunk via the nose or mouth. This was a popular method of consumption among shamans, as the strength of the brew could be adjusted to deliver the massive doses they preferred. The provenance of the tobacco used in making tea was a matter of great importance. For instance, Acawaio men would travel to a special stream to collect 'Mountain Spirit' tobacco, which was steeped in the water of the stream to enhance its potency. Drinking tobacco also presented the opportunity of mixing other narcotics into the brew. Novice shamans would sometimes add a dash of the fluids they collected from a dead shaman, and a qualified shaman's tea was often loaded with other hallucinogenic plant extracts. Tobacco was drunk in sufficient quantities at shamanic initiation ceremonies to induce vomiting, paralysis and, occasionally, death. Even everyday tobacco drinkers attributed mystic powers to their brew. Hunters of the Mashco tribe drank to communicate with the game animals that they wished to kill. Hunters in some tribes would apply tobacco juice as eye drops in order to help them see in the dark. In several cases this privilege was extended to their hunting dogs.

Tobacco tea was also 'drunk' via the anus where it was introduced in the form of a clyster, using a hollow length of cane or bone, or with a bulb made out of animal skin and a bone or reed nozzle. An early example of such a device, dating from AD 500, has been discovered in the tomb of a Colombian shaman. Tobacco enemas were used for both medicinal and spiritual purposes. The Aguarana tribe, for example, employed enemas to protect apprentice shamans from were-jaguars during initiation ceremonies. A further variant of tobacco drinking, tobacco licking, was popular among some South American civilizations. This form of consumption involved boiling down tobacco tea into a syrup or a jelly known as ambil. Sometimes alkaline salts were added, and the syrup thickened with manioc starch. Ambil was used by dipping a stick or finger into the jelly and rubbing it over the gums. It was often carried in a little pot on a string around its devotee's neck.

Far more widespread than tobacco licking, and uniquely American, was tobacco sniffing. Snuff was prepared by drying, toasting then pulverizing cured tobacco leaves, and the resultant powder was blended then stored in calabashes or bottle gourds. Other plants were often snuffed in conjunction with tobacco, especially coca. In the days before paper currency, insufflators (snuffing machines) were created in a variety of shapes. The most simple of these consisted of a hollow reed or bone, which was inserted into a single nostril. An equally common design was Y-shaped, which enabled the snuffer to accelerate the charge by blowing down one tube, with the other up one nostril, or to take both barrels at once. Some tribes blew snuff up one another's noses using elongated insufflators to speed up the snuff's narcotic effect. Snuffing was the preferred method of tobacco consumption of the Incas, whose remarkable civilization, governed by semi-divine rulers along communist lines, and distinguished by an impressive road-building programme, was exterminated by the Spaniards in the sixteenth century.

The most common form of tobacco consumption in South America was smoking, usually using cigars or a simple form of cigarette consisting of cured strips of tobacco wrapped in musa leaves or corn husks. The act of smoking was not merely a method of tobacco consumption, but an integral part of ritual. Shamans used tobacco smoke for healing and blessing, and also as a form of food to nourish their guiding spirits. Shamans believed that they entered into a contract with the spirit world upon initiation, whereby they undertook to provide sustenance to the spirits in the form of tobacco, in return for receiving healing and other powers. Spirits that had taken up residence within the shaman's body were nourished by the tobacco he himself used, whereas those living in crystals or other sacred objects had smoke blown over them. For example, the shaman of the Campa tribe owned a sacred rock which he would smoke over and 'feed' daily with tobacco juice.

The preferred implement for smoking tobacco was the cigar, which could be of prodigious size, especially those prepared by shamans, where examples of a metre or more in length are not uncommon. These were made from rolls of cured tobacco, often wrapped around a stick or the rib of a banana leaf. Some tribes developed special cigar supports, resembling giant tuning forks, which could be held in the hand, or whose sharp end could be stuck in the ground to support these monsters. Shamans' cigars occasionally were sprinkled with *carana* granules which affected the vocal cords and masked the voice of the smoker, giving it a harsh, deep inflection which was considered appropriate for ritual discourse between mankind and the spiritual powers.

Cigars for hedonistic or everyday use came in an immense variety of shapes and sizes. They were the common currency of the tobacco habit in South America and, in addition to providing pleasure to their users, served simple social roles. Smoking was acknowledged to alleviate hunger – a useful attribute for subsistence level societies. Cigars were offered as tokens of welcome and friendship. They were smoked for relaxation and as self-administered medicine. They were employed to keep evil spirits and thunderstorms at bay. Such was their ubiquity in South American society that it is impossible to isolate a single or prime reason for smoking. The question 'Why smoke?' could have been answered effectively and truthfully with 'Because we are humans.'

Following tobacco's historical progress from its centre of origin northwards into Central America, methods of consumption became less diverse, with smoking gaining at the expense of other tobacco habits. The earliest historical record of tobacco use in Central America resides among the artefacts of the Mayans, a sophisticated metropolitan civilization that flourished between about 2000 BC and AD 900.

The Mayans farmed tobacco and considered its consumption to be not only a form of pleasure, but also a ritual of immense significance. At least two of their principal gods were habitual smokers.

The Mayans celebrated tobacco's place in their culture. Numerous extant artefacts testify to their official and recreational dedication to the weed. Its ritual importance derived from its symbolic role as a medium of transporting the blood offerings which were central to Mayan theology and culture towards heaven. The Mayans believed mankind had been created from the blood of God, and so humanity's purpose on Earth was to give blood back as frequently as possible, whether other people's or their own. In addition to tearing out the still-beating hearts of sacrificial victims, the Mayans obtained blood by pulling thorn studded ropes through their own tongues or by flaying their penises. The blood thus shed to nourish and communicate with their gods was then soaked up in paper or strips of cloth and burned. The smoke it released was a vehicle which conveyed the Mayans' offerings to their deities. When not indulging in devout self-mutilation, the Mayans would pray by smoking, which represented communication with the divine, an association it had enjoyed to the south. Mayans also smoked tobacco with no greater purpose in mind than hedonism. Whilst smoking had a solemn ritual function, it was also an exercise in pleasure.

The figure below shows a Mayan noble enjoying a cigar, seated on a jaguar skin. This prince of the royal blood reaches forward towards a vision serpent whose head can be seen emerging from the shell at his feet. The image is carved into a section of conch shell and the object is not known to fulfil a specific ritual function, which suggests it was a personal treasure and depicts a private moment. This early evidence of the use of tobacco for relaxation and contemplation demonstrates its importance in pre-Columbian America as a leisure activity. The Mayans were the world's first historians of smoking and their elegant depictions of smokers, both mythical and real, speak of their devotion to tobacco. Their civilization flourished in isolation for nearly three millennia

A Mayan noble meditates

until it fell apart in the early tenth century AD. By the time Europeans came across their ruins in the jungles of Central America the great cities of the Mayans had been deserted for 500 years.

Subsequent to the Mayans, another great tobacco-using American civilization, the Aztecs, flourished in what is now Mexico. The Aztecs, like the Mayans, smoked for health, ritual and pleasure. They were imperialists, expanding their domains through conquest, and had inherited the Mayans' fascination with feathers, tobacco and blood. They were also herbalists, using tobacco to cure a number of diseases. Some of these ancient prescriptions survive in a codex transcribed by their Spanish conquerors. They are veritable witches' brews with perhaps more magical than active ingredients. For instance, the Aztec cure for gout involved a tobacco tea foot bath, using tobacco leaves that had first been left in a ditch where ants could walk on them, followed by 'serpentine rabbits' ground into a powder, a small white or red stone, a yellow flint and 'the flesh and excrement of a fox, which you must burn to a crisp'. Tobacco also played less positive roles among the Aztecs. It was considered a necessary accompaniment to the ceremonies at which thousands of captives were slain in sacrifice to the god Tezcatlipoca. This association of the weed with blood and war had travelled with it from South America, where warriors were smoked over before battle, and were said to be driven into a 'demoniac fury' by this form of passive smoking.

Archaeological evidence, in the form of a primitive pipe, indicates that tobacco had reached the northern part of the American continent prior to 2500 BC. Its prehistorical use appears to have been near universal, from the swamps and deserts in the south, through the forests and across the great plains to the limits of tree growth in the north. With the exception of the frozen tundra of Alaska and Canada, wherever there were men, tobacco was consumed. Some tribes who practised no other form of agriculture planted and cared for tobacco. The Tlingit Indians, an Alaskan tribe of hunter-gatherers, took a break from hunting and gathering to cultivate tobacco. Similarly some of the plains tribes, including the Blackfoot and the Crow, to whom growing vegetables was anathema, planted and nourished the weed.

Smoking was a defining habit of the diverse tribes and civilizations that occupied pre-Columbian North America. Every one of its cultures, living and vanished, used tobacco. In some cases, the only mementoes civilizations have left to posterity have been their smoking apparatus. Not only was tobacco use common to all the inhabitants of North America, but they seem to have been unanimous in their selection of the pipe for its enjoyment. This focus resulted in an accumulation of mythology around tobacco pipes, so that they came to serve distinct social and ritual functions from the tobacco that they burned. The pre-eminence of pipes also led to smoking becoming an activity restricted by sex. Pipes were for men, and by the time of the Europeans' arrival in North America, tobacco use in many tribes was a male preserve.

The best archaeological record of ancient North American tobacco use is found in the relics of the mound-building cultures who inhabited an area stretching from the lower Mississippi valley to New York. Three successive civilizations, named the Adena, Hopewell and Mississippi, constructed immense, zoomorphic burial mounds in which the possessions of the dead were stored with their bodies for use in the afterlife. This and other aspects of their culture, including pipe decoration with animal species native to Central but not North America, their pottery design and consumption of maize, suggest links to the Mayans.

These Mayans in exile commenced mound building in the lower Mississippi valley around 1500 BC. Over the next millennium or so they spread north into the Ohio River valley and their civilization entered what is known as its Adena period. They continued to expand north, reaching Manhattan Island in about 300 AD, where they farmed corn and tobacco on a large scale. At this stage their culture is called the Hopewell, and had achieved a considerable degree of complexity. The Adena/Hopewell were sophisticated craftsmen, who made objects for aesthetic as well as utilitarian purposes. They excelled in copper work and stone carving, and their creative expression was at its most acute when making tobacco pipes. Many thousands of these beautiful artefacts have been discovered in burial mounds. Collectively, they represent the greatest repository of early North American artistry. Pipes were treasured possessions and were buried beside their owners. When a man died, his pipe went with him.

Tobacco pipe design evolved through three stages among the Adena/Hopewell culture, during which decoration became more intricate, while design grew more functional. Early pipes were large, heavy objects carved from a single piece of stone. Their bowls arose from the middle of an arched stone base, one end of which the smoker would hold while they inhaled from the other. These pipes were too cumbersome to be easily portable and it appears that their smoking was a collective, sedentary and ritual occupation. However, as craftsmanship evolved, both the form and function of the tobacco pipe changed. Pipes became smaller, and were decorated with exquisite carvings of animals and men. These pipes were for personal use: when an individual wished to communicate with his totem, or the spirit world, he would smoke his tobacco pipe, which had been shaped to represent his totemic animal or a recognized messenger with the spirits. Smoking appears to have been a form of profound meditation – a device to raise the smoker above the distractions of a world of flesh. As a smoker inhaled he literally drank the substance of the eternal, and the smoke he exhaled in turn represented his questions or desires transubstantiated into a form acceptable to the spirits. As soon as he lit his pipe, the smoker would exhale once in the four cardinal directions – north, south, east then west – in order to orientate his prayers towards their intended recipients.

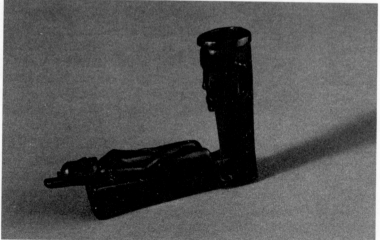

Pipe dreams

Birds were the most common animals carved on Adena/Hopewell pipe bowls because of their ability to travel through the air, which was conceived of as a separate world. Ducks and other waterfowl were particularly popular because they were able to travel through all of the worlds: air, water and land. For similar reasons beavers were frequently carved on bowls, as were representations of the prince of Mayan mythology, the frog, which being a creature of both land and water, was likewise a spirit messenger. Human heads were sometimes depicted, as ever, carved to face the smoker. These pipes may have been used when smokers wished to communicate with a particular member of the deceased.

From about AD 700 onwards, the Adena/Hopewell culture entered an epoch of retreat and decline during which it is known as the Mississippian period. The Mississippians continued to smoke and to bury pipes in mounds but a regression in design had occurred. Pipes had become impracticably large – too heavy for personal use, and were collectively employed in 'meditation groups', instead of solacing solitary smokers. The usual subjects carved on their bowls were human figures, as if the Mississippians were trying ever harder to communicate with their ancestors and their race's past glories. Their history ended in 1731 when the tribe's remnants were annihilated by the French.

A more complete record remains of the tobacco habits of the plains tribes of North America, who survived longer after the coming of the white man. Although illiterate, their oral legends were chronicled by their white replacements who, smokers themselves, were particularly sympathetic to those concerning tobacco. As a consequence, records exist of not just how but why these early Americans smoked. Many of these justifications were identical to those employed in South and Central America. Clearly, the weed had been accompanied on its journey north by the existing reasons for its use. These included its familiar role as a medicine, although the number of diseases it was believed to cure had diminished. However, it was still used to fight toothache and to heal snake bites, and was also recommended for cleaning the lungs.

Tobacco's integration into the cultures of various North American tribes stretches back to their creation myths, where the weed was usually represented as a divine and precious gift. Some myths depicted tobacco, or its spirit, as a beautiful celestial maiden. This attribution of femininity was later adopted by the white man to help make sense of his fondness for tobacco. Tobacco had also accumulated some new ritual functions, reflecting its importance in the relatively drug impoverished north of the continent. Amongst many tribes, smoking was never practised as an end in itself but always for some definite purpose forming part of the day's routine, or as a rite prescribed by tribal custom. Whenever important decisions concerning the family, the tribe or inter-tribal relations were required, a meeting of braves would be convened. The matters at hand would be discussed while a pipe was passed around. In this manner, the tobacco pipe served a similar function to the Speaker's Mace in Britain's parliament – it was a ritual prop, the means of transforming argument into debate.

Tobacco pipes form a significant part of the cultural legacy of native Americans. Enormous quantities of time and energy were dedicated to their manufacture and decoration. Amongst the plains tribesmen, pipes were often the only possession that they had not strictly necessary to the procurement or storage of sustenance – the only non-utensil. The plains tribesman committed as much passion to the manufacture of his pipe as any Renaissance craftsman to the creation of a chalice. The fine carving of their pipe bowls gives evidence of the reverence with which the ritual and paraphernalia of smoking were held. An astonishing legend surrounded the source of stone used in the pipe bowls of many different and widely separated tribes: that somewhere in America's heart a special pipe mine existed, provided by their common god for every man to use, and at which all inter-tribal hostilities had to be suspended. So deep were the taboos surrounding this fabled pipe mine that it was not until 1840 that the first white man set eyes on it.

On occasions, tribes used natural instead of artificial means to create their pipes. For instance, in the northern Californian area of Klamath, where smoking was a 'cult', the faddish Karok Indians invented an eco-friendly method of making pipe stems from arrow-wood by enlisting the aid of the salmon weevil, a parasitic grub that infected their principal foodstuff. Arrow-wood possesses a spongy pith which would be soaked in salmon oil, before introducing a weevil to one end of the stem, which was left to eat its way out to the other, feeding on the delicious fish oils which constituted its usual diet. The result was a hollow tube which was attached to a stone pipe bowl. In contrast to this complex interaction with insect istars, the Pueblo Indians, who inhabited the area around Santa Fe, smoked their tobacco through simple pottery tubes, fired from the same adobe clay with which they clad their houses.

Tobacco pipes *per se* had several important public roles of their own in Indian societies: 'the mythology surrounding the pipe was no less cosmic than that of tobacco itself'. Not only the weed, but its chariots had formal duties. Pipes were used to seal oaths, to declare war, and to provide a safe conduct. The Omaha Indians who inhabited most of Oklahoma had a pair of war pipes and a pair of peace pipes. The former were plain, the latter highly ornamented with feathers and hair tufts, whose arrangement 'was charged with significance, and the peace pipe of any particular tribe was as easily recognizable to other tribes as was the banner or the coat of arms of a feudal Lord . . . the pipe often served as a pass or safe-conduct for a messenger through hostile territory'. The use of 'peace pipes' extended tobacco's South American associations with friendship, so that sharing a pipe was a symbol of amity across the continent, and a prelude to all interactions between strangers.

Many tribes did not conceive of their pipes as weapons requiring specific ammunition. They were exuberantly fond of fumigating their lungs. They seem to have possessed two separate habits – tobacco and smoking. A surprising diversity of flora was burned in their pipes. These other plants were used as flavour enhancers for tobacco, for mystical reasons, and, in times of shortage, as tobacco substitutes. Some tribes could not function without smoking, as their social rituals had been constructed around it. Few tribes could meet in peace unless a pipe was present. Commerce would have been impossible without tobacco as it was the only widely traded material on the continent. Even if a tribe had run out of tobacco, it still had to smoke. In some tobacco smoking tribes the rituals surrounding the practice, when taken in sum, amounted to a smoking religion.

Isolated from the rest of their species for almost 18,000 years, the peoples of the Americas had matched or exceeded their overseas counterparts against many measures of cultural achievements. They had alphabets, pyramids and calendars. Their botanical knowledge and agricultural skills provided them with the resources to feed the great cities that they established. The modern world is awash with daily reminders of these peoples' ingenuity, discoveries and art. In addition to terrestrial locations, some of the universe's visible constellations still bear their names. And their most popular gift to the rest of humanity has since become its most common habit. Did the first American smoker ever dream of the places and times in which he would be imitated, of the men and women who would one day share his pleasure?

2

Confrontation

*The race for China – Christopher Columbus
discovers America, and tobacco – the first
European smokers – their reactions to
tobacco – smoking and Satan – smoking
and syphilis – the Spanish conquest of
South America – the first prohibition*

Tobacco, and indeed the American continent, came to the
outside world's attention by accident. From about 1419
onwards, Europeans had taken to making long Atlantic
voyages. The Portuguese led the field. In 1439, they established a colony in the Azores. In 1456, they occupied the
Cape Verde Islands. By 1488, they had explored the west
coast of Africa as far as Cape Town and were poised to
make their debut on the Indian Ocean. The goal of these
voyages was China, known as 'Cathay', a semi-mythical
country which was the home of silk, pearls, porcelain and
spices. The best known European authority on China was
Marco Polo and his partly imaginary diary of his travels
was the most comprehensive account of the country that the
Western world possessed. Trade with China was carried out
overland through the territories of the infidel then across
the Mediterranean in the fleets of the Genoese and Venetians.
Since the trade consisted principally of valuable and desirable commodities, and Constantinople, a hub of the traditional route, had fallen to the hostile Turks in 1453, northern

Europeans were increasingly obsessed with discovering a maritime route to China.

Contrary to popular misconception, fifteenth-century Europeans did not believe the world to be flat. Contemporary science agreed it was round or nearly so. Christopher Columbus, a Genoese sea captain, thought it was breast shaped: '(it) does not have the kind of sphericity described by the authorities, but that it has the shape of a pear . . . or that it is as if one had a very round ball, on part of which something like a woman's teat were placed'. Driven by his theory of a mammary globe, Columbus sought a royal patron to enable him to reach China, and its eastern treasures, by sailing west. He needed this sanction because at that time, as a private individual, he could only lay claim to whatever land he might discover with the backing of a king or queen. He applied first to King John II of Portugal but was ignored. The Portuguese were busy exploiting Africa and setting up a slave trade.

However, in neighbouring Spain, Christian conquistadores were cleansing their own country of infidels, finishing a war that had occupied them continuously for over 700 years. Columbus's appearance at the Spanish court in 1492, within weeks of the surrender of Granada, the last heathen stronghold, was opportune; the bold venture he suggested entirely in keeping with the martial yet pious atmosphere then prevalent in Spain. After Granada, why not China? Columbus was granted the money and ships he needed to give his plan to get to China by travelling the wrong way a try.

Columbus set out from Palos in Spain on 3 August 1492, with a fleet of three ships, his passport – a heraldic scroll identifying him as a royal servant and requesting the assistance and protection of foreign powers – and personalised letters from Ferdinand and Isabella, Spain's 'Catholic Kings', addressed to, among others, the Great Khan of China. He reached Las Palmas in the Canary Islands the same month and on 6 September, cast off and headed west. Columbus chose the right stretch of ocean at the wrong time of year. Had he waited two months, the tradewinds would have settled in and whisked him over the Atlantic. Instead, he wallowed along, surrounded by the most featureless horizon in the northern hemisphere. That part of the Atlantic is relatively barren. There are few birds, fewer dolphins, and it is outside the path of whale migration routes. Even the sea bed far below is uninteresting – an oceanographer's nightmare composed largely of abyssal plains.

As Columbus progressed, the heavens were invaded by unfamiliar constellations. The Southern Cross rose upon the shoulder of Orion. Showers of shooting stars descended from the luminous tropical skies. Lightning bolts connected overhanging clouds to the sea. When the air was clear strange birds were seen, which were taken by his crew to be a sign that land was close. Sadly these were tropical birds which, despite their tiny beaks and plumed tail feathers, venture as far from land as albatrosses, indeed, only visit land once a year to breed. Columbus, notwithstanding his geographical misconceptions, really had embarked on a voyage into the unknown. The best information available on what lay across the Atlantic was contained in Greek legends, or had been kept a secret.

On 11 October 1492, his crew near mutinous, his provisions exhausted, Columbus saw a small moving light on the horizon. The next day he reached an island in the Bahamas. He dressed in silk, went ashore, named the island San Salvador and claimed it in the names of Ferdinand and Isabella of Spain. He was met by members of the local tribe, who, sensing the importance of the occasion, offered him gifts of beads, fruit and dried leaves. In return, Columbus gave them two red hats, which were known to be popular with the Chinese. Columbus had brought with him, to act as an interpreter, a converted Jew who spoke Hebrew, Chaldean and Arabic, which 'eastern' languages were presumed to be useful in China. This man was set to work and it was revealed that a larger island lay westwards where there were 'gold, spices, and great ships and merchants'. Convinced this must be Hangchow, a major oriental city, Columbus weighed anchor. The natives' confusing gift of dried leaves was thrown overboard for the fish – the white men did not yet know the value of tobacco.

Columbus reached Cuba on 28 October. Upon arrival he sent an expedition inland, carrying the royal letters of introduction, in search of the Great Khan. The expedition returned with the letters undelivered but reported having seen many oddities which included meeting a number of natives 'with a little lighted brand made from a kind of plant whose aroma it was their custom to inhale'. Two of its members, Rodrigo de Jerez and the interpreter Luis de Torres, had even gone so far as to try the custom for themselves, thus becoming the first Europeans to smoke tobacco.

This first encounter with tobacco merits close examination. It was, after a fashion, Europe's first kiss with the American continent – the first contact with something really new. What did Europeans make of smoking? They had no precedent for the act. No one smoked anything in Europe. They burned things to produce sweet smells, to sniff, but not to inhale. Smoke was for dispersal, not consumption. Without domestic precedents, the Europeans even lacked the vocabulary to describe smoking. And the Spanish were a very special sort of European. They had spent the last seven centuries at war within their own country, a struggle they pursued in the twin names of religion and liberty. They obtained their morality from the Bible or fairy tales and their determination from constant battle. The Spaniards were famously cruel to infidels and notoriously suspicious of their habits.

Sadly, the first European smokers' reactions were not documented, or if they were, were not transcribed, for the actual log of Columbus's first voyage, with its historic entry of 'Land Ho!', has been lost. There is a copy, made in 1514 by Fra Bartolomé de Las Casas, later Saint Bartolomé, who did not believe unreservedly that Columbus's discoveries had been a good thing. St Bartolomé re-wrote Columbus's log in the third person and added his own commentary, which included observations he himself made when he eventually travelled to the Americas. This is what he had to say about the Old World's first encounter with smoking:

> These two Christians [Rodrigo de Jerez and Luis de
> Torres] met many people on the road, men and women,
> and the men always with a firebrand in their hands,
> and certain herbs to take their smokes which are some
> dried herbs, put in a certain leaf, dry also, after the
> fashion of a musket made of paper, such as boys make
> on the feast of the Holy Ghost, lit at one end and at
> the other they chew or suck and take in with their
> breath that smoke which dulls their flesh and as it were
> intoxicates and so they say that they do not feel weari-
> ness. Those muskets or whatever we call them they call
> tobacos.

The impression conveyed is one of wonder, the similes used,
true to St Bartolomé's profession, ecclesiastical. His description focuses on tobacco's ability to intoxicate, and notes the
Indians' claim that it dulls their appetite. No mention,
however, is made of tobacco's sensuous qualities. What did
it smell like? What did it taste like? It clearly inebriated
Indians, but what effect would it have on a Christian? The
commentary also neglects to consider why the Indians were
smoking, beyond noting their claim that it suppressed their
hunger and enhanced their endurance. The oversight seems
surprising in the face of such novel behaviour. The Americans
had some very good answers to this question: they smoked
for their health, for their gods and for the pleasurable side
effects. But they could only hint at these impulses across the
language barrier.

Both of the original European smokers, Rodrigo de Jerez
and Luis de Torres, are reputed to have become habitual
smokers in the course of the three months for which
Columbus's little fleet lingered in the 'Indies' before setting
sail for Spain. The voyage home was tempestuous, and
Columbus did not reach Spain until March 1493. He was
fêted at the Spanish court upon his arrival, and was granted
a coat of arms of great splendour. He returned to the Indies
in September 1493 with seventeen ships and 1200 men. It
was easy to find recruits for this second voyage. Among

other curiosities, Columbus had brought back a little gold. Although not even enough to cover the cost of his expedition, its very existence in the New World was a sufficient spur to many adventurers to make the voyage. Hernán Cortés, the conqueror of Mexico, summed up his contemporaries' lust for the yellow stuff: 'I and my companions suffer from a disease of the heart that can only be cured by gold.'

The Spaniards returned to the Caribbean islands with the intention of taking absolute possession of them. Their claims to the American continent were settled by Pope Alexander VI, coincidentally a Spaniard, who divided ownership of the globe between the Spanish and Portuguese in 1494. The Spaniards had also evolved a legal justification for their policy of enslaving or exterminating any humans they might come across in the New World. This was enshrined in a document named the *requeriemento*, which pretended to derive authority from the Bible's Book of Joshua and which was to be read aloud to natives before an attack might be made on them. The *requeriemento* commenced with a brief history of the world since its creation, proceeded to the establishment of the Holy Roman Catholic Church, provided the genealogy of Spain's monarchs, and concluded with a demand that the listeners renounce, if different, their faith, culture and freedom, or: 'We will do all the harm and damage that we can . . . and we protest that the deaths and losses which shall accrue from this are your fault.'

The Spanish conquistadores were rigorous in their observation of the *requeriemento*, ensuring lawyers were present to note its proclamation everywhere they went, though it was often read in Spanish 'to the trees'. In addition to this legal weapon, they had the advantages of surprise, infection and technology. The Spaniards found their conquest of the Caribbean islands a remarkably easy task, so easy that within twenty years of Columbus's first landfall they were lamenting an unforeseen consequence of the original inhabitants' extermination – a shortage of labour. The Spaniards had had to do little fighting during this period. Their bodies were hosts to numerous fatal diseases against which the Americans had no immunity, and often the mere presence of Spaniards was enough to depopulate an island.

Despite such limitations, sufficient Indians remained as slaves to enable the Spaniards to observe them smoking and to revise their initial impressions of tobacco. By now, familiarity had bred contempt and the Christians condemned the habit. Gonzalo Fernandez de Oviedo, military governor of Hispaniola, had this to say of smoking:

> Among other evil practices, the Indians have one that is especially harmful: the ingestion of a certain kind of smoke they call tobacco, in order to produce a state of stupor. Their chiefs employ a tube shaped like a Y, filled with the lighted weed, inserting the forked extremities into their nostrils . . . In this way they imbibe the smoke until they become unconscious and lie sprawling on the ground like men in a drunken slumber . . .

When Bartolomé de Las Casas travelled to the New World he was dismayed to find Europeans taking tobacco. In the opinion of a saint, tobacco was a vice: 'I have known Spaniards in this isle of Hispaniola who were wont to take [tobacco], and being reproved for it and told that it was a vice, they replied that it was not in their power to stop taking [it].' Note that St Bartolomé alludes to another of tobacco's novel properties: it was strangely compelling. Whilst the Spaniards could observe the consequences of smoking, there was a factor in its consumption that they could not understand, and that was addiction. The concept of addiction was unknown in Europe. Excessive or obsessive indulgence in anything venal was categorized plainly and simply as a sin.

Relations between the Spaniards and Americans were limited to exploitation and sex. Despite frequent exhortations from Spain and missionaries like St Bartolomé to treat the 'Indians' kindly and to attempt their conversion to the Catholic faith, the conquistadores kept them as slaves or toys, and, in suitably biblical retribution, the Indians gave them syphilis. Syphilis provided an impetus to Spanish tobacco use that hitherto had been missing. The Spaniards had noticed that Americans used tobacco as medicine against the disease and began to do the same themselves. As Oviedo observed:

> I am aware that some Christians have already adopted the habit, especially those who have contracted syphilis, for they say that in the state of ecstasy caused by the smoke they no longer feel their pain. In my opinion the man who acts thus merely passes while still alive into a deathly stupor. It seems to me that it would be better to suffer the pain, which they make their excuse, for it is certain that smoking will never cure the disease.

Luckless sufferers of syphilis aside, who were understood to deserve their fate after succumbing to the sin of lust, the Christians developed a spiritual aversion to tobacco, which was exacerbated by the discovery of its satanic functions amongst the Indians they despised. The weed's reputation as an active tool of the Antichrist, serving as his messenger to his human servants, developed when the Spaniards ventured on to the mainland of the American continent.

In northern Venezuela, Oviedo detected tobacco's use by shamans as a tool for communication with the arch-fiend.

> They venerate and dread the devil very much, and the [shamans] say they can see him and have seen him many times . . . These [shamans] are their priests and in every important town there is a [shaman] to whom everyone goes to ask what is going to happen, whether it will rain, or whether the year will be dry or abundant, or whether they should go to war against their enemies, or refrain from doing so, or whether the Christians are well disposed, or will kill them, or finally, they ask all they wish to know. And the [shaman] says he will reply after having a consultation with the devil.

And in order to have the consultation he shuts himself
into a cabin alone, and there he makes use of . . .
tobaccos . . . And as soon as he comes out he says this
is what the devil tells him, answering the questions
which have been asked, according to the desires of those
whom he wishes to satisfy.

Oviedo further observed that every tribesman looked on
tobacco as an oracle, and if in doubt whether to 'fish or to
plant or to know if he should hunt or if his wife loves him
each one is his own prophet, since, having twisted the leaves
of this herb together in a roll to the size of an ear of corn,
they light it at one end, and they hold it in their mouth
while it burns.' After smoking half of his cigar the tribesman
would throw it on the ground and make his decision in
accordance with the shape that the extinguished stub
assumed.

The gradual hardening of the Spaniards' attitude to
tobacco reflected the increasing opposition they were
meeting to their conquests. Juan de Grijalva, coincidentally
the first European to smoke on continental American soil,
was repulsed by its inhabitants after he abused their hospitality and gifts of tobacco. Juan Ponce de León, leader of
an expedition to Florida, was fatally wounded and carried
back to Cuba to die. The object of his voyage was fascinating, if ironic. In addition to the Garden of Eden, Paradise
and various other wonders of ancient times, America was
suspected of harbouring the Fountain of Youth, a spring or
pond that was believed to bestow eternal youth on anyone
who drank its waters. Ponce de León went to Florida in
search of immortality and his example has been followed,
with equal success, by America's senior citizens ever since.

Ponce de León's defeat and demise raised the Spaniards'
estimation of the Indians in every aspect bar the spiritual.
Instead of being godless savages, they were now perceived
as savage warriors, actively in league with Satan and his
minions. The opinion was reinforced when Hernán Cortés
set out from Cuba in 1519 with the aim of subduing and

settling Mexico. Cortés's expedition was the largest yet to attempt the mainland, consisting of 508 soldiers and 16 horses. The soldiers were easily recruited in Cuba where the supply of Indian slaves had run out and the alternative to conquest was manual labour. Their opponents were the Aztecs, the great metropolitan power whose capital is now Mexico City. The Aztecs had built their empire through conquest, but they were unpopular rulers in the lands they occupied on account of their appetite for sacrificial victims, especially children and virgins, who were ceremoniously abducted by Aztec warriors dressed in colourful plumage before their hearts were ripped out and used as offerings.

Cortés exploited the discontent among the Aztecs' subject races and although his army never contained more than 700 Spaniards, he was assisted in his battles by up to 150,000 local allies. He fought his way towards the Aztec capital where he was civilly received by the Aztecs' ruler, Montezuma II. Where lesser men might have been apprehensive of their circumstances – surrounded by hundreds of thousands of hostile warriors whose actions were directed by priests who relied on human sacrifice as a tool for decision-making – Cortés saw only opportunity, and besides, the city was full of gold.

Cortés kidnapped Montezuma and held him to ransom for all his treasure. Whilst Montezuma was a hostage of the Spaniards in his own palace, he was permitted to retain his entourage and to continue his domestic rituals, including indulging in his favourite post-prandial habit of smoking. Montezuma, on account of his status as supreme ruler and descendant of the sun itself, dined alone, sitting on a low stool 'soft and richly worked' behind a little golden screen. There were compensations for this formal isolation. Four beautiful maidens served him a choice of 300 dishes while hunchbacks and dwarfs gambolled around his feet telling jokes. Most importantly: 'there were also placed on the table three tubes much painted and gilded which held . . . certain herbs which they call tobacco and when he had finished eating, after they had danced before him and sung and the table was removed, he inhaled the smoke from one of those tubes'.

The harmony portrayed in the first European description of a smoke after dinner did not last. The Spaniards wanted the Aztecs' treasure, their country, and possibly their souls, although the eagerness with which they separated these last from their bodies suggests the contrary. Spanish Catholics had been brought up with a very clear idea of the Devil and his worshippers, and the Aztecs held a surprising similarity in the flesh to the Satanists of Spanish myth. Their idols bore a coincidental resemblance to Spanish representations of the arch-fiend and their penchant for theatrical human sacrifices closed the minds of Cortés's soldiers to mercy, who witnessed a number of their own countrymen suffer this horrible fate. When the conquistadores overthrew the temples at the summits of Aztec pyramids they found chambers clotted with rotting human blood nearly a foot thick, tended by priests bleeding from self-inflicted wounds to their tongues and ears, whose hair was a stinking mass of coagulated blood, and whose persons were surrounded by flies. The conquistadores gained their final victory over the infidel on 'the thirteenth day of August at the time of Vespers on the Saint Day of St Hipolito', 1521.

Tobacco suffered by association with Aztec religion, and its use does not appear to have been adopted by Cortes's followers. However, while the Spaniards were winning Mexico for Jesus and their king, other Europeans were visiting the New World, and were making their own assessments of tobacco. Unlike the Spaniards, these voyagers did not travel with the express intentions of conquest and conversion, and sought to co-operate with the Indians they encountered instead of claiming them. As a consequence, they were more curious about American customs. The earliest European account of what it feels like to smoke, as opposed to what it looks like, or what it does to Indians, is contained in the *Brief Receipt* of the Breton seafarer, Jacques Cartier, who travelled to the New World in 1534, inspired by the treasure fellow Bretons had seized from one of Cortes's ships on its way back to Spain. Cartier came across tobacco in what is now Canada, where the Iroquois tribe offered him a pipe upon arrival. Cartier was clearly fascinated by the habit, for his account of smoking summarized how tobacco was grown, how it was prepared, and why it was taken, as well as what it is like to experience smoking:

> There groweth also a certain kind of herb, whereof in summer they make great provision for all the year, making great account of it, and only men use of it, and first, they cause it to be dried in the sun, then wear it about their neck wrapped in a little beast's skin made like a little bag with a hollow piece of stone or wood like a pipe: then when they please they make a powder of it, and then put it in one of the ends of the said cornet or pipe, and laying a coal of fire upon it, at the other end suck so long that they fill their bodies full of smoke, till that it cometh out of their mouth and nostrils, even as out of the tunnel of a chimney. They say that this does keep them warm and in health. We ourselves have tried the same smoke, and having put it in our mouths, it seemed that they had filled it with pepper dust it is so hot.

The words chosen by Cartier to describe tobacco's taste, hot and peppery, reflect his need to express the habit in the alimentary terms that Europeans might understand. His observations were made without judgement, and as such might be read as the first European endorsement of the weed. According to Cartier, smoking was an interesting habit with practical purposes.

In addition to smoking, the Spaniards and other New World voyagers had come across snuffing, the method of drug ingestion unique to the New World. According to Columbus's son, the Great Admiral had observed tobacco snuffing on one of his visits, and noted its associations with religion. In the house where the Indians kept their idols was 'a finely wrought table, round like a wooden dish in which is some powder which is placed by them on the heads of these [idols] in performing a certain ceremony; then with a cane that has two branches which they place in their nostrils, they snuff up this dust . . . with this powder they lose consciousness and become like drunken men'.

Snuffing was also documented by an Italian visitor to Brazil, Amerigo Vespucci, who further observed the natives indulging in another drug habit which, had he not been so supercilious, might have made him a fortune:

> We found . . . the most brutish and uncivilized people . . . each had his cheeks bulging with a certain green herb which they chewed like cattle . . . and hanging from his neck each carried two dried gourds one of which was full of the herb he kept in his mouth, the other full of a certain white flour like powdered chalk. Frequently, each put a small stick . . . into the gourd filled with flour, then drew it forth and put it in both sides of his cheeks, thus mixing the flour with the herb which their mouths contained.

The 'herb' in this case was not tobacco but coca leaves, the powder lime to release the active alkaloid, i.e. cocaine. Whilst Amerigo missed the opportunity to become the first international cocaine dealer, he had the consolation of having the new continent named after him: America.

The discovery of America gave rise to an orgy of naming – a New World, unknown plants and novel animals had been found, all requiring identities. It is comparable to the twentieth century when the mass of inventions that appeared, such as television, RAM and automobiles similarly expanded the vocabulary. The origin of the name 'tobacco' has been a matter of etymological debate. Its use among early commentators on the New World was not universal. They had touched upon the Atlantic coastline of the Americas in places thousands of miles apart, where they had come into contact with tribes of vastly different cultures and languages who did not know of each other's existence, and who employed a variety of names for the weed. The Europeans seldom understood what the natives were saying and this confusion is evident in one of the explanations of the etymology of tobacco – that it derives from the Cuban 'dattukupa', which means 'we are smoking'. Whatever its meaning may be, 'tobacco' is almost certainly Caribbean in origin and represents the European phoneticization of an untraced native word. Whilst the weed was only a curiosity, whose entire method of consumption had to be described from start to finish, several rival names flourished, including 'cohoba' and 'petun'. The latter appears in the writings of a lapsed Carmelite friar, André Thevet, who visited Brazil in the hope of establishing a Protestant presence in the New World: 'There is another secret herb which they name in their language Petun, the which they most commonly bear about them for that they esteem it marvellously profitable for many things.' But clarity of name is essential for trade between nations of different tongues. As soon as it had a commercial value 'tobacco' was adopted as the plant's identity and Thevet's rival 'petun' was attached to tobacco's cousin the petunia flower as a consolation – charm instead of harm.

Thevet, like Cartier, tried the weed himself and left the first record of the effects of smoking on the white man: 'The Christians that do now inhabit [Brazil] are become very desirous of this herb and its perfume, although that the first use thereof is not without danger, before that one is accustomed thereto, for this smoke causes sweats and weakness, even to fall into a syncope.' Interestingly, he attributes the compulsion tobacco creates in its users to its scent, as if this in itself was sufficient reason to take up smoking. Other European commentators did not find the smell so appealing. Girolamo Benzoni, who travelled through Central America in the 1540s, hated tobacco smoke, even at second hand: 'It is hardly possible for one who has not experienced it to realize how injurious, how poisonous is this hellish practice; many a time when travelling through Guatemala, if I chanced

to enter the dwelling of an Indian who was in the habit of using the herb – tobacco – the Mexicans call it – I was compelled to fly as soon as the diabolical stench reached my nostrils.'

Whilst other nations were making forays along the Atlantic coast of the New World, the Spaniards had discovered and subdued another empire equal in size and strength to that of the Aztecs. Their next victims, the Incas, occupied what is now Peru, and their domains were at their greatest in extent when the Spaniards arrived, although recently divided by a civil war. Francisco Pizarro, conqueror of the Incas, imitated his fellow countryman Cortés in seeking a meeting with the Inca ruler, took him hostage, then burned him at the stake in his own barracks.

The Spaniards treated the Incas' civilization in a similar manner to the Aztecs'. While they kept their roads they destroyed or depopulated the Incas' cities, prohibited their religion and enslaved the remnants of their subjects. Peru had 9 million people in 1532, in 1570 only 1.5 million of these remained alive. The Spaniards even slaughtered the Incas' beast of burden, the llama. Llamas had been the Incas' haulage system, assembled in convoys of 500 or even a thousand, laden with back-packs then marched 'in perfect order and in obedience to the voice of the driver'. These charming and spirited little animals were used to transport dried tobacco and coca leaves to the distant palaces of the Incas. The coca was chewed, the tobacco snuffed and used, unlike coca, for strictly medicinal purposes. The llamas were decimated by the conquistadores who killed them, surprisingly, for their brains which they turned into a stew, discarding the remainder of the animal.

A measure of the changes wrought by the Spaniards in the New World can be gained from the account of one of Cortés's lieutenants, comparing his first sight of Mexico to its state a few years later:

> We were amazed and said that it was like the *Enchantments of Amadis* [a romance that was popular reading among the conquistadores] on account of the great towers and pyramids and buildings rising from the water . . . I stood looking at it and thought that never in the world would there be discovered other lands such as these, for at that time there was no Peru, nor any thought of it. Today, of all these wonders that I then beheld, all is overthrown and lost.

When one civilization absorbs another via conquest, it sometimes adopts its victims' culture to the extent that it is later hard to distinguish the victor from the vanquished. In such circumstances the vanquished entity may be said to have triumphed – through the perpetuation of its gods, its art, or its pleasures. The Aztecs and the Incas may have been destroyed by the Spaniards but their tobacco habits have been adopted since by the entire world. If an Aztec and a Roman were transported to the twenty-first century the

Aztec might be the less mystified. He would know why people were smoking.

Tobacco use finally made inroads into the Spanish population of the Americas via the Roman Catholic clergy, who developed a fondness for snuff. Spain's religious orders, following the efforts of St Bartolomé, had scrutinized the Indians' beliefs in the hope that they might discover a simple way to explain Christianity to them and win their voluntary conversion. It appears that in the course of their studies the clergy resolved that tobacco, at least in the form of snuff, was compatible with their creed, and adopted the satanic messenger they had previously condemned. The clergy's recognition of snuff reflects their acceptance of the Indians' tendency to use tobacco to communicate with the divine, and snuff was less of a distraction during mass than smoking. However, the habit soon spread from the congregation to the altar, resulting in the first tobacco prohibition in history. An ecclesiastical decree issued in Lima in 1588 ordained that 'It is forbidden under penalty of eternal damnation for priests, about to administer the sacraments, either to take the smoke of . . . tobacco into the mouth, or the powder of tobacco into the nose, even under the guise of medicine, before the service of the mass.'

By the middle of the sixteenth century Spain's American adventures were paying dividends of such prodigious size that its new found wealth threatened to alter the balance of power in Europe. Silver flowed from its Peruvian mines in such quantities that it devalued the holdings of every other Old World nation. This impoverishment inspired the efforts of other European countries in the Americas, and a general interest in the treasures that the new found world contained.

3

Transmission

*Tobacco arrives in Europe where it is planted
in palace gardens – its medical potential and
its use as a cure for cancer – English pirates
and smoking for pleasure – tobacco makes its
West End debut and is proclaimed as a muse
– the English attempt plantations in the New
World – the weed travels around the world*

Rumours of tobacco reached Europe before the plant itself.
Columbus did not bring any back on any of his voyages,
nor did the other Europeans who followed him across the
Atlantic. The chances of a mere curiosity reaching Old World
shores were slim, as were the chances of any single ship
getting to, and back from, the New World intact. As a consequence, Europe learned of tobacco's existence second hand.
Descriptions of the plant and of the curious habit of smoking
were featured in many accounts of the marvels of the New
World, but they attracted less attention than the rumours
of Fountains of Youth and rivers of gold that these reports
also contained. In addition, the few published descriptions
of the act of smoking had done little for its reputation. Its
satanic associations in Oviedo's work have been noted, and
its popularity amongst Indians further diminished its appeal
to Christians.

Without tobacco users to observe indulging, the Europeans
could not formulate a clear idea of this new herb, and

whether it should be sought or scorned. Eyewitness accounts at their disposal concurred that tobacco was sniffed or smoked, that it intoxicated natives who claimed to take it as an appetite suppressant, and that white men with venereal disease had started smoking.

It seems that Europeans were unable to imagine tobacco consumption from these descriptions alone, and rather than being prepared for the worst, reacted with horror and consternation when returning sailors smoked. Whilst travellers to the New World had set out expecting to see wonders, their counterparts at home were less open minded and did not welcome surprises. Besides, their Christian culture associated smoke with the Devil, who was sometimes represented in pictures as a miasma emanating from various human orifices. The hapless Rodrigo de Jerez, the first ever European smoker, was reputed to have been imprisoned by the Inquisition on his return to Spain for public smoking, and detained in a dungeon for three years – presumably, the time it took to exorcize this novel form of possession.

However, accounts of the tobacco plant's medical properties had excited the Europeans' curiosity, which triumphed over their spiritual revulsion, and tobacco seeds were carried back to Spain and Portugal in the 1550s. Tobacco began its European life in palace gardens where it was studied and nurtured by court physicians. This initial association with royalty greatly helped its reputation. Courtiers were servile imitators and scrambled for cuttings to plant in their own residences. Kings were jealous of each other's novelties and their ambassadors were instructed to obtain tobacco seeds for their masters' courts. One such, Jean Nicot, sent from France to Lisbon in 1559 to arrange a marriage between the fifteen-year-old son of the Portuguese king and the sixteen-year-old daughter of Henri II of France, begged some tobacco cuttings from Damiao de Goes, the celebrated Portuguese botanist, which he raised in the gardens of the French embassy. Thus far, tobacco had been cultivated principally on account of its beauty. It flourished in European soil and spread to many gardens on the strength of its appearance alone. This had fed back into its yet untested medical reputation, thus quickening the plant's dispersal, for even if, as some sceptics claimed, it had no value as a cure, no one would object to its charming presence in their flower beds. Nicot, however, had other plans for his tobacco plants.

The problem with tobacco was that it could not be identified in European herbals, which depicted only plants grown around the Mediterranean, and therefore it was impossible to know how tobacco was to be used, or what it was meant to cure. European medicine of the period was based on the concept of 'humours', formulated by Galen (AD 131-201), a Greek doctor who practised in the declining years of the Roman empire. Galen's system viewed illnesses as disturbances arising in the natural balance of human body fluids which were known collectively as *humours*. Four in number, the humours consisted of phlegm, yellow bile, black bile and blood. Each humour had an essence – *wet* or *dry* – and was either *hot* or *cold*. Blood, for instance, was *hot* and *wet*. Health was restored by rebalancing the humours, which was achieved by diet, medicine and blood letting. Although it differed to the South American concept of illness, which focused on spiritual instead of mechanical causes, Galenism was no more advanced in its ability to treat diseases, and relied equally heavily on herbal remedies. If tobacco could be found a place in the Galenic system – identified as *hot* and *dry*, for example, in which case it could be used to counter an excess of phlegm which was *cold* and *wet* – then its medical value would become apparent. This was the breakthrough Jean Nicot sought when he started experimenting with the weed.

Some of the rumours concerning tobacco's reputed healing powers had highlighted its potential as a cure for cancer, and Nicot determined to put the herb to the test. When he chanced upon a Lisbon man with a tumour he treated him with an ointment made from tobacco leaves and effected a complete cure. After further and equally

successful experiments, Nicot was confident enough to send
plants and seeds to the Queen of France, Catherine de'
Medici, whose fascination for alchemy, magic and superstition made her the perfect recipient of a novel and potent
herb. Nicot also sent a letter containing testimonials of
tobacco's powers and later advised that in addition to being
applied externally, it might be taken as snuff. In the imitative atmosphere that prevailed in the French court, people
without wounds or tumours began to take tobacco-snuff,
or the 'Nicotian Herb' as they called it, as a preventative,
in a similar spirit to the consumption of vitamins today.
They found the habit strangely compulsive, and tobacco use
began to spread as quickly as the plant itself.

In Portugal meanwhile, Nicot's experiments had become
the talk of Lisbon. The Papal Nuncio, Monsignor Prospero
St Croce, obtained some seeds for the Pope that His Holiness
instructed to be sown in the gardens of the Vatican. Tobacco
also reached Italy under the auspices of Cosimo de' Medici,
Grand Duke of Tuscany and the richest man in Europe, who
obtained his seeds from his relative, the Queen of France.
They were entrusted to the Bishop of Saluzzo, who nurtured
them in his palace. From here monks carried the seeds to
other Italian kingdoms, and from thence tobacco was taken
to Bohemia. It is ironic that a number of eminent Christians
were charged with the care of the same plant whose
consumption had been linked with the arch-fiend and enemy
of their church.

In 1565, tobacco's medical reputation received a significant boost when Nicolás Monardes, a doctor of Seville,
published a pamphlet listing the herb's healing properties.
His eulogy, entitled *Joyful News of our Newe Founde Worlde*
has never, even in the golden age of cigarette advertising,
been surpassed in enthusiasm or excess. *Joyful News*
commenced with an exposition of tobacco's benign effects
upon the human brain, which it cleansed and invigorated.
The pamphlet then proceeded to recommend tobacco for
'Griefs of the breast . . . rottenness at the mouth, and for
them that are short of wind.' Warming to his task, Dr

Monardes next revealed tobacco to be an effective cure for any illness of any internal organ, for bad breath, especially in children who have eaten too much meat, for kidney stones, for tape worms, for toothache, for tiger bites and for wounds from poison arrows, indeed for 'any other manner of wound'. Having established a case for tobacco as a universal cure for humans, Monardes noted its efficacy when used on animals. Tobacco, he asserted, could heal cattle of 'new wounds and rotten', of maggot infections, of foot-and-mouth disease, and of any parasite that had or might trouble them. In a final arabesque, *Joyful News* returned to human patients and gave a case history of tobacco curing dandruff, even after 'masters of surgery had done their diligence – all to no profit'.

Joyful News was also careful to refute some of the negative accusations concerning the plant that had emerged from the New World, including the charge that tobacco was the Devil's herb. Monardes admitted the demonic connection, but only on the basis that so was every other herb, including parsley and rhubarb. He also, by way of turning an accusation to his advantage, acknowledged that tobacco had the useful satanic qualities of suppressing hunger, curing insomnia and enhancing endurance. It is not recorded whether Monardes gave up his medical practice to become a tobacconist.

The publication of *Joyful News* provoked a wave of interest in tobacco across Europe. The pamphlet was translated into Latin, English, French and Italian. Its French translator, anxious to ensure maximum sales, exploited the royal link and dedicated the book to Catherine de' Medici. Its frontispiece, an engraving of a tobacco plant, was captioned 'Herba Medici' in a flattering attempt to name the plant after the queen. Tobacco's portrait also appeared in the herbals of the period, which were guides to plants with culinary or medicinal properties. One such (*Nova stirpium adversaria*, published in Antwerp in 1576) accompanied its picture of the tobacco plant with a depiction of smoking, which habit was still so rare on the continent that whilst the engravers

Home grown – tobacco cultivated in Europe with a 'fundibulo' inset

could draw a tobacco plant from life, they had to resort to their imagination for a likeness of the smoker.

At this stage in tobacco's European existence, most tobacco consumed on the continent was produced domestically and much of it was used, like other herbs, in the form of poultices. However, the snuffing habit was spreading beyond the French court, principally through the efforts of the country's physicians. Small quantities of cured tobacco were imported from Cuba to meet this demand, whose stronger 'medicinal' qualities led to its preference over home-grown amongst court health fanatics. The appetite for imported tobacco grew quickly and it became a regular feature in the cargoes of Spanish and Portuguese vessels. The retail price of tobacco, whatever its source, was very high – only the rich could afford such an infallible medicine. Demand rose nonetheless, as the result of a strangely cumulative desire amongst tobacco users. Step by step, tobacco's cultivation in and transportation from the New World was becoming an international business.

Tobacco use followed the plant into the Lowlands, into the Italian and German states, and even into Switzerland. A country's reaction to the introduction of tobacco was highly revealing of its national character. The French used it to ward off illness and preserve beauty; in the Italian states, it was entrusted to the care of priests and administered on their advice; in the German principalities it was examined in accordance with the best current scientific principles and declared a 'violent herb'; in Switzerland it was tested first on a dog before being recommended for human consumption. One European country, however, focused on an entirely different justification for tobacco use: pleasure.

There is often a portion of a nation's history on which its inhabitants reflect with pride, which the chroniclers of other nations, employing a less partial eye, might rather describe as a period of shame. Such is the age of the Elizabethan sea knights, whose names still stir English hearts – Sir John Hawkins, Sir Walter Ralegh, Sir Francis Drake, whose daring adventures, or piratical raids, established their fame, ruined England's international reputation in the more traditional realms of continental diplomacy, and popularized smoking in the Old World.

After the death in 1558 of Queen Mary, wife of Philip II of Spain, the English had been in a state of war with that country. Much of the hostilities took place at sea, off the coasts of Europe and the Americas, where English ships preyed on the Spanish galleons. The English did not yet have a presence in the New World, and found it easier to rob a galleon than a nation. Unlike the Incas, the English did not fear the Spanish: 'the Spaniards have been noted to be of small dispatch: *Mi venga la muerte de Spagna*: Let my death come from Spain; for then it will be sure to be long in coming.' Besides, their ships were faster than the Spaniards', and if an engagement fared badly they could and did run away. When the popular heroes of these encounters returned to England, they carried tobacco as well as treasure.

Hawkins, Drake and Ralegh have all been credited with introducing tobacco use to England. While the legends surrounding Sir Walter Ralegh are the most numerous, including the favourite of many school-books – that he was dowsed by a servant who observed him smoking and believed him to be aflame – Ralegh was not the first Englishman to smoke on his native soil.

The most likely candidates were members of Admiral Hawkins's slave-trading cum piratical voyage around the Caribbean in 1562. Hawkins recorded Floridian Indians smoking and his men may have adopted their habit. The English were taking a different approach to the New World from their Spanish adversaries. There were no great golden empires left to subdue, and the Indians they encountered they sought to befriend instead of enslave. As a consequence, they were more inclined to try out than condemn their customs. Also, being less idol-minded than the Catholic Spanish – most of England's religious art having been destroyed in the bout of iconoclasm that followed the Reformation – they did not associate smoke so intimately with Satan.

Sir Francis Drake was probably the next man to bring tobacco to England. He had encountered the weed on his circumnavigation of the globe between 1577 and 1580. On 5 June 1579 he dropped anchor on the Pacific coast of America near what is now San Francisco, where he was met by native Americans whose houses were close to the water's edge. These natives were not only friendly, but subservient: 'when they came unto us, they greatly wondered at the things that we brought, but our general [Drake] (according to his natural and accustomed humanity) courteously entreated them, and liberally bestowed on them necessary things to cover their nakedness, whereupon they supposed us to be gods, and would not be persuaded to the contrary'. The natives brought Drake presents suitable for gods which consisted of 'feathers and bags of tobacco'. Their king then ceded his peoples and land to Drake who accepted them in the name of Queen Elizabeth of England and named California *Nova Albion*. He set up a monument of title, engraving a post and nailing up a sixpenny piece. The tobacco stayed on board Drake's ship as he struck out across the Pacific, towards England, home and glory.

Ralegh's involvement with tobacco dates from the return of his first Virginian expedition in 1586. Its mariners had taken up smoking, and had also brought back a pair of 'poor barbourous savages' who had been supplied with tobacco to last them through their exile. William Camden (1551–1623), a contemporary historian, reported that 'These men who were thus brought back were the first that I know of that brought into England that Indian plant which they call Tabacca and Nicotia, or Tobacco.'

But Camden was wrong. Tobacco was certainly in use in England by 1571 when Pierre Pena and Mathias de L'Obel described the plant and its consumption via combustion with a degree of familiarity: 'You will observe shipmasters and all others who come back from out there [the New World] using little funnels, made of palm leaves or straw, in the extreme end of which they stuff crumbled dried leaves of this plant. This they light, and opening their mouths as much as they can, they suck in the smoke with their breath. By this they say their hunger and thirst are allayed, their strength restored and their spirits refreshed.'

Smoking spread rapidly throughout England, and the English adopted the pipe as their implement of choice for their new habit. The English preference for a pipe over the cigar usually employed on the Spanish Main is explained by the smoking practices of the Indians they had made contact with in North America, who had been refining the art of pipe smoking for thousands of years. The first pipes the English used were small clay devices that were usually the property of the 'tabagies' or smoking dens that were springing up all over the country to cater for the aficionados of the weed. Their domestic use was first noted in Harrison's *Chronology* (1573): 'In these days the taking in of the smoke of the Indian herb called Tabaco, by an instrument formed like a little ladle, wherebye it passeth from the mouth into the head and stomach is greatly taken up and used in England.'

As the English sailors had acquired their tobacco habit and its mode of execution from a different ethnic source to the Spaniards, it came with different rituals attached, and the English evolved their own justifications for its use. A number of these were medical – in particular the weed's ability to cause its user to expel excess phlegm, usually by expectoration, which was then considered a valuable benefit to the inhabitants of dank climates. Other rationales for smoking focused on tobacco's intoxicating powers – a desirable quality for a substance in Elizabethan England, where the stupor tobacco was reputed to cause was observed with the keenest interest. Pena and L'Obel commented that smokers 'declare also that their brain is lulled by a pleasing drunkenness'. The intrepid herbalists even experimented with the weed on themselves: 'We have found, nevertheless, that while we drink this in, it does not inebriate quickly, nor drive one mad . . . but it fills the ventricles of the brain with a certain vapourous perfume.' Smoking became known as 'dry drinking' and its fearless practitioners as 'tobacconists'. Tobacco in the Elizabethan age was also called 'sotweed', after its ability to turn its user into an incoherent fool. However, most Englishmen preferred stupefaction to health, and the answer an Elizabethan would have given to the eternal question 'Why smoke?' would have been 'For pleasure.'

Smoking soon developed from a curiosity into a craze. Much of its popularity, especially in the higher strata of English society, can be attributed to Sir Walter Ralegh, who was handsome, virile and eloquent, and whose mannerisms were widely imitated. At the court of Queen Elizabeth I smoking was welcomed as symbolic of the spirit of adventure its proponents personified. Even the Virgin Queen herself, a balding spinster with bad teeth, was persuaded by the charms of her favourite Ralegh to inhale a little of the fumes of the same weed her continental counterparts were smearing on tumours or sniffing to relieve the 'vapours' with which they claimed to be afflicted.

Once smoking had become *de rigueur* at court, it proved irresistible to English society. Elizabethans quickly invented rituals to accompany the new vice. These soon matched in sophistication the ancient rites developed by Americans. Smokers were nicknamed 'reeking gallants' and would attend the theatre weighed down by pipes and the other paraphernalia of their pleasure. The average gallant required so many smoking accessories, including tobacco boxes, knives, tongs and pipes, that a dedicated manservant was needed to carry them. The act of expelling smoke was not in itself enough, so smoking was embellished with flamboyant behaviour, and became a kind of parlour game. Pipes were lit from a coal passed from smoker to smoker on the point of a sword. 'Tobacconists' were expected to possess a lengthy repertoire of tricks involving a combination of facial contortions and the expulsion of carefully shaped tobacco clouds. A new art evolved, dedicated to the perfect smoke ring.

By the closing years of the sixteenth century smoking had spread throughout England with such rapidity that visiting foreigners remarked upon the habit as a local curiosity. In 1598, Paul Hentzner, a German visitor to the bear-baiting pit in Southwark commented:

At all these spectacles, and everywhere else, the English are constantly smoking the Nicotian weed . . . and generally in this manner: They have pipes on purpose made of clay, into the further end of which they put the herb, so dry that it may be rubbed into powder, and lighting it, they draw the smoke into their mouths, which they puff out again through their nostrils, like funnels, along with it plenty of phlegm and defluxion from the head.

The smoking plague attracted criticism as well as the wonder
of foreigners. The weed had its English detractors, including
the historian Camden who observed tobacco's proliferation
with distaste. 'In a short time many men, everywhere, some
from wantonness, some for health sake, with an insatiable
desire and greediness, sucked in the stinking smoke thereof,
through an earthen pipe, which presently they blew out again
through their nostrils, insomuch that tobacco shops are now
as ordinary as taverns and tap houses.'

Smoking commenced its remarkable associations with
English literature and European culture in the Elizabethan
age. Christopher Marlowe, sexual adventurer, spy, hyper-
bolic playwright, author of *Dr Faustus*, that perfect depiction of compromise and damnation, declared: 'All they that
love not boys and tobacco are fools', and this maxim seems
to have been taken to heart by the circle of great writers
of which he was a part. Tobacco, and smoking, began to
appear in the poetry and plays of the time. Smoking's most
vociferous proponent was Ben Jonson, who satirized the
habit in his play *Every Man in his Humour* (1598), thus
providing a valuable contemporary analysis of the new
fashion. Tobacco's powers are expounded by the character
of Captain Bobadill, who reveals he has lived on tobacco
and nothing else in the New World. 'I have been in the
Indies, where this herb grows, where neither myself, nor a
dozen gentlemen more, of my knowledge, have received the
taste of any other nutriment in the world, for the space of
one and twenty weeks, but the fume of this simple only:
therefore, it cannot be, but 'tis most divine!' Captain
Bobadill proceeds to enumerate tobacco's curative powers
and concludes that it is 'the most sovereign and precious
weed, that ever the Earth tendered to the use of man' (Act
3, scene 2).

Some of the then current objections to smoking were
voiced in the same play by the character of Cob, a simpleton:
'I marvel what pleasure or felicity they have in taking this
roguish tobacco! It's good for nothing but to choke a man,
and fill him full of smoke, and embers: there were four died

out of the house, last week, with taking of it, and two more the bell went for, yestemight; one of them, they say, will never escape it: he voided a bushel of soot yesterday, upward and downward.' Cob's fears highlight the superstition that smoking left a carbonized residue, like any other fire. The image of a smoker choked and constipated with soot was to persist in England for many years, and, for some of this time, was taken to be a medical fact as opposed to a playwright's invention.

Tobacco returned to centre stage in 1599 in the sequel to *Every Man in his Humour – Every Man out of his Humour*. Smoking was depicted in the new work as an essential accomplishment of the aspiring gallant and satirical examples of gallant behaviour were provided. The play features private smoking lessons, which guarantee students will learn 'the most gentlemanlike use of tobacco: as first, to give it the most exquisite perfume; then, to know all the delicate sweet forms for the assumption of it: as also the rare corollary, and practice of the Cubanebullition, Euripus and Whiff, which he shall receive, or take in, here in London, and evapourate at Uxbridge, or further, if it please him'. Presumably, holding one's smoke was considered as important as the ability to drink excessively without losing the command of reason.

Tobacco's reach extended beyond England's playwrights to its poets and philosophers. Spenser hailed 'divine tobacco' in his *Faerie Queene*, an allegorical poem dedicated to the greatness of Elizabeth's reign. Samuel Rowlands celebrated the weed's nourishing qualities in a collection of light verse entitled *Knave of Clubs*:

> But he's a frugal man indeed,
> That with a leaf can dine,
> And needs no napkin for his hands,
> His fingers' ends to wipe,
> But keeps his kitchen in a box,
> And roast meat in a pipe.

However, the philosopher Francis Bacon drew attention to another aspect of smoking – its apparent ability to control its users: 'In our time the use of tobacco is growing greatly and conquers men with a certain secret pleasure, so that those who have once become accustomed thereto can later hardly be restrained therefrom.' As in so many other instances, Bacon was a man ahead of his time, yet he lacked the vocabulary to scrutinize this aspect of tobacco consumption closely. The Elizabethans did not possess an equivalent concept to modern addiction. Although they were masters of excess, the notion that a human with an immortal soul could be rendered the helpless slave of a vegetable would have appeared ridiculous to them.

There was a single, glorious exception to the chorus of praise for tobacco in Elizabethan literature: no record exists of William Shakespeare smoking, nor is there any reference to the habit, or tobacco, in any of his work. Shakespeare was an admirer of Marlowe and a friend of Jonson's, acting in many of his plays including *Every Man in his Humour*, but whether Shakespeare's silence on the subject was a reproach to his profligate fellow playwrights, or if he considered smoking merely to be a fad, too ephemeral to merit comment, and unsuitable for characters like King Lear, Henry V or Julius Caesar, the grocer's son kept his writing a smoke-free zone. There is, however, archaeological evidence proving that smoking was going on around the Shakespeare household in Stratford-upon-Avon during his life, and fingers have been pointed at the bard. Recent investigations have uncovered a number of clay pipes from Shakespeare's time, whose bowls contain the combustion residues of tobacco, cannabis, an unidentified hallucinogen and cocaine.

England's infatuation with tobacco was marred by a single factor – its cost. The English did not yet have plantations in the Indies, and had to rely for their tobacco supply on capture, trade with foreigners or smuggling. By 1586, its value was so great that Sir Francis Drake brought back a ship-load after sacking the Spanish settlements of

Santo Domingo, Cartagena and San Agustín, in the knowledge that the weed was as good as cash. The clay pipes that have survived from the Elizabethan age are known by collectors as 'elfin' or 'faerie' pipes on account of their tiny size, and are a testament to the expense of tobacco. By 1598, tobacco cost £4 10s. per pound. By way of comparison, a mug of ale cost a penny and a young whore with good teeth was less than a shilling a throw. Even the most influential men in the land were having to squeeze their contacts in order to assume a healthy supply of the weed, as the following letter from a City of London trader to Sir Robert Cecil demonstrates: 'According to your request, I have sent the greatest part of my store of tobaca by the bearer, wishing that the same might be to your good liking. But this tobaca I have had this six months, which was such as my son brought home, but since that time I have had none. At this present there is none that is good to be had for money.'

Part of the shortage was solved by the introduction of home-grown tobacco. Cultivation had begun at Winchcombe, in the heart of England, by 1590. Sir Walter Ralegh planted tobacco at his Irish estate in Cork, where, unlike his other New World import (potatoes), it gained immediate popularity among the locals, who are reported to have stolen the tobacco, but to have pelted the estate manager with potatoes when he suggested that these alien tubers might be fit for human consumption.

Despite the cost, tobacco use kept rising. England's first official import statistics, from 1602, show that 16,128 pounds of foreign weed entered the Port of London – a significant quantity of so valuable a commodity. Although 'tabagies' had sprung up all over England, the bulk of the tobacco trade was controlled by apothecaries, who imported tobacco as a medicine. In order to preserve the mystick secrets of their trade, and to increase profits, the apothecaries adulterated tobacco which led to a public outcry. In 1595 the first work dedicated to tobacco was published, whose purpose appears to have been to stir up resentment against the men of medicine. Its author, Anthony Chute, noted the existence of a 'smoakie Societie' that had banded together so that 'hoarding apothecaries might be glad to abate their prices of their mingle mangle which forsooth they will not sell under unreasonable rate, when there is scarce good to be got'.

Public concerns over adulteration were reflected by Ben Jonson in his masterpiece *The Alchemist*, and the dubious quality of some of the tobacco on sale – 'mingle mangle' reeking of stale wine and soiled underclothing – led to a minor backlash against the Elizabethan smoking craze. Once the issue of adulteration had been raised, it was seized upon by a new sect of Christians named Puritans, who considered the body to be God's temple and therefore not lightly to be defiled. In 1602, the first English anti-smoking tract was published, entitled *Worke for Chimney Sweepers*, which resurrected some of the Spanish objections to the vice of savages:

> But hence thou Pagan Idol: tawny weed.
> Come not within our Fairie Coasts to feed,
> Our wit-worn gallants, with the scent of thee,
> Sent for the devil and his Company,
> Go charm the priest and Indian cannibals,
> That ceremoniously dead sleeping falls
> Flat on the ground, by virtue of thy scent.

Such doggerel, however, did nothing to extinguish demand and could not solve the problems of supply. England was the largest market for tobacco in Europe, yet the English lacked even a safe harbour on the vast American coast. It was not until a century after Columbus's discovery that they affected to pay any attention to settling the New World at all. This delay can be explained in part by the state of war that existed between Spain and England. It was easier to steal than settle, so English resources were allocated to military ventures instead of colonies which could not be expected to be profitable at once, and in the interim would be targets for Spanish reprisals. There were also many practical obstacles: the New World was not only far away, but hard to find. Sixteenth-century ships were small and unseaworthy, and navigation was an art, not a science.

To ignorance can be added the perennial obsession of English politicians with domestic policy. Overseas was unimportant and interesting the English public or administration in anywhere other than England was an uphill task. Every Englishman who had been to the New World, including Sir Francis Drake and Sir Walter Ralegh, argued, in vain, that it was in the national interest to establish some form of presence there, even if only a naval base from which the Spaniards might be plundered more efficiently. Despite Drake claiming California for Queen Elizabeth, two centuries were to pass before another English voice was heard around San Francisco. Richard Hakluyt, a contemporary historian, summed up the adventurers' frustration: 'I marvel not a little that since the first discovery of America (which is now full fourscore and ten years) after so great conquest and planting by the Spaniards and the Portuguese there, that we of England could never have the grace to set fast footing in such fertile and temperate places as are left as yet unpossessed by them.'

The case for an American colony, and therefore independent tobacco supply, was represented by heroes and statesmen to no avail. In the end they were forced to resort to advertising. In 1587, Sir Walter Ralegh commissioned Thomas Harriot, the surveyor of an experimental settlement on a part of America's east coast, that had been named Virginia in honour of Queen Elizabeth's presumed state of innocence, to brush up the New World's image. The resulting sales brochure, 'A Brief and True Report of the New Founde Lande of Virginia', was intended to entice Englishmen overseas. It painted a picture of a fertile paradise, with a superabundance of both the familiar and the exotic. The New World's climate was perfect, its forests were full of fruits and game, its rivers overflowed with fish. Not least of all, its inhabitants were friendly and helpful.

Virginia, in short, was the land of opportunity. Harriot's report contained a eulogy to tobacco which the amicable Indians used to purge their heads, clear their skin, and cleanse their bodies 'from obstructions', with the result that the Indians were 'notably preserved in health, & know not many grievous diseases wherewithall we in England are often afflicted'.

Harriot also addressed the thorny issue of tobacco's spiritual functions amongst savages, but instead of depicting them as Satanists, smoking only to communicate with the prince of darkness, his account emphasized the Indians' childlike simplicity and the gratification they derived from their favourite weed. Tobacco was

> of so precious estimation amongst them that they think their Gods are marvellously delighted therewith: Whereupon they sometimes make hallowed fires & cast some of the powder therein for a sacrifice: if there is a storm on the waters, to pacifie their gods, they cast some up into the air and into the water . . . also, after an escape from danger, they cast some into the air like-wise: but all done with strange gestures, stamping, sometimes dancing, clapping of hands, holding up of hands and staring up into the heavens, uttering there-withall and chattering strange words and noises.

Harriot's prose was intended to convey the natural excitement the Americans associated with tobacco. His bias was confirmed when he admitted to trying the herb himself: 'we ourselves during the time we were there used to suck it after their manner, as also since our return and have found many rare and wonderful experiences of the virtues thereof . . . the use of it by so many of late, men and women of great calling as else, and some learned Physicians also, is sufficient witness'.

Sadly, life in Virginia did not measure up to Harriot's descriptions. There was little opportunity to smoke at leisure, and notwithstanding the rude health and plenty enjoyed by the Indians, Englishmen tended to fall ill or to starve. The English attempted two American colonies in the 1580s, both situated in Roanoke in the Chesapeake Bay, both of which quickly failed. The first of these was sponsored by Sir Walter Ralegh, as part of a fact-finding, settling and piracy mission.

New friends

One hundred and seven men were deposited on shore, then the English fleet set out to make something from the voyage. It succeeded in this aim by capturing a Spanish galleon, the *Santa Maria*, on the journey home. When Francis Drake dropped by Roanoke two years later, only a handful of the colonists were still alive. At the sight of his ship they begged to be taken back to England.

The second English attempt at a colony, also in Roanoke, suffered an even more tragic fate. It was founded in 1587, with 144 men, women and children who realised very quickly that they needed more resources from England. Unfortunately, their relief ship was diverted to fight the Spanish Armada, and despite the pleas of John White, the colony's governor, who had left his wife and daughter Virginia (the first person of English blood to be born on American soil) in the colony, a return voyage was forbidden. When a relief mission was finally permitted in 1590 the second Roanoke colony had vanished without trace. Its fate has yet to be determined.

Despite such setbacks, Ralegh never lost faith in the need for colonies or the power of advertising. His account of his disastrous trip to Guyana (1595) was entitled *The Discoverie of the Large, Rich and Beautiful Empire of Guiana, with a Relation of the Great and Golden City of Manoa (which the Spaniards Call Eldorado)*. A man before his time, Ralegh fell from favour after Queen Elizabeth's death in 1603 and was beheaded in 1618. Contemporaneous accounts claim that his ultimate action on this earth was to smoke a final pipe of his beloved tobacco. His pipe case is preserved in London's Wallace Collection. It bears the following inscription, which, as will be shown, summarizes the services performed by tobacco for the early English colonists: '*Comes meus fuit illo miserrimo tempo*' ('It was my companion in that most wretched time').

Whilst the English were learning to smoke, their sailors and those of other European nations spread tobacco, as they had done so many diseases, worldwide. They took a ritual from one race and transplanted it as an unexplained habit to many other civilizations. The weed reached the Levant and Middle East via the fleets of the Genoese and the Venetians, whilst the Spanish and the Portuguese introduced smoking to Africa and Asia. Tobacco arrived in most of these countries as a seaman's habit. It was a substance to be used for pleasure and was not burdened with any other coherent explanations for its consumption. It was a motherless child, which gave its new admirers the opportunity to invent justifications for their impulses to smoke. Not all of them bothered. In some cases there was little analysis, for mankind is ever eager to embrace new kinds of stimulation.

The Japanese, who had received tobacco courtesy of a shipwreck in 1542, took to the weed with the same unthinking gusto as the English. Samurai knights formed smoking clubs and commissioned elegant paraphernalia in a manner reminiscent of the Elizabethan 'reeking gallants'. They favoured ornate silver tobacco pipes which they strapped to their backs or tucked into their kimonos beside their legendary swords. Tobacco enchanted the upper strata of Japanese society, and even received imperial approval: a gift of tobacco seeds made to the shogun in 1596 was sowed and raised in the gardens of his palace in Kyoto.

Japanese smokers also resembled their English counterparts in their matter of fact approach to smoking. A contemporary description of the habit's genesis in Japan, written by Saka, a doctor of Nagasaki, is little more than an algorithm: 'of late a new thing has come into fashion called "tobacco", it consists of large leaves which are cut up and of which one drinks the smoke'. The description does not even mention the property of tobacco which had been a matter of universal comment in early European accounts – its ability to suppress the appetite. It is possible that in a country motivated by Zen and the spirit of minimalism, an antidote to gluttony was not a priority. In Japan, tobacco had been abstracted from the rituals of and reasons for its consumption, and it flourished in their absence.

Tobacco arrived in many other Asian countries via the

trans-Pacific trade route that the Spanish had opened up between Mexico and the Philippines in 1571. From the Philippines it was dispersed to Thailand, Cambodia, Sri Lanka and Goa. From Goa its use spread rapidly throughout the Indian subcontinent, so that within a few years of its introduction an English ambassador to the court of the Great Mogul was able to report: 'They sow tobacco in vast plenty and smoke it much.'

Unlike Columbus, tobacco reached China, probably via the Portuguese trading enclave of Macao. The Chinese took up tobacco initially as protection against malaria. Chang Chieh Pin, a physician from Shanyin, recorded its appearance in the territory of the Celestial Emperor:

> Inquiring for the beginnings of tobacco smoking, we find that it is connected with the subjugation of Yunan province. When our forces entered this malaria infested region, almost everyone was infected by this disease with the exception of a single battalion. To the question why they had kept well, these men replied that they all indulged in tobacco. For this reason it was diffused into all parts of the country. Everyone in the South West, old and young without exception, is at present . . . smoking by day and night.

Tobacco's adoption by the Chinese resulted in its incorporation into a third system of healing. Chinese medicine was based on the assumption of the unity of creation which was reflected by a group of ordering characteristics, through which all natural objects could be identified and understood. The first level of definition consisted of two 'I': *yin* and *yang*, opposing yet complementary forces, one passive and female, the other active and male, one associated with darkness and earth, the other with lightness and the heavens. Illness was caused by an imbalance between yin and yang, and cures were effected by restoring the equilibrium. The concept of balancing forces was not dissimilar to the Western system of humours and tobacco was identified as belonging to yin, i.e. it was *hot*. It was accepted as a cure for troubles due to cold and moisture – an identical diagnosis to those reached in Europe under Galenic theory. It was also believed to be an excellent aid to digestion 'for this reason many people use it as a substitute for wine and tea, and never get tired of it, even when smoking all day long'. Tobacco became a favourite amongst Chinese poets, who smoked to assist the act of composition. Yau Lu, a sixteenth-century poet, whilst acknowledging that incautious use of the weed 'could make one tipsy', celebrated its inspirational properties as the 'gold-silk smoke'.

The continent on which tobacco had the most profound initial impact was Africa. After Vasco da Gama's historic circumnavigation of the globe in 1497–9, the Portuguese had taken control of the ancient Arab trading stations in Sofala, Mombasa and Hormuz. Tobacco, which it seemed the Koran had overlooked, was welcomed by Muslims as a permitted drug and distributed by their dhows up and down Africa's east coast. The Portuguese also created their own chain of trading posts on the continent's west coast where they introduced tobacco to every tribe that they encountered. These in turn passed the weed on to other tribes so that tobacco reached the heart of the dark continent centuries before the white man. Initially, supplies were imported from the Portuguese colony of Brazil, but by the end of the sixteenth century tobacco was being cultivated throughout equatorial Africa. It was carried in canoes, along forest paths, acquiring different names as it passed between tribes, shedding or gaining properties so that some were introduced to the weed as a miracle worker, while at other destinations it arrived in the guise of an aphrodisiac.

Africans evolved their own creation myths for tobacco – often involving its discovery by wanderers, or its seizure from the gods themselves in a Promethean act of daring. These embellishments, typical to oral cultures, simultaneously provided a justification for smoking tobacco so that it became an act with specific purposes. For example, the

Bushong tribe of the Congo attribute tobacco's origin to
Lusana Lumunbala, a Bushong prodigal, who returned to
his tribe after twenty years of wandering. Entering his native
village at sunset, he was met by a group of old men, the
contemporaries of his circumcision ceremony. None of these
prematurely aged playfellows recognized Lusana Lumunbala,
but when he told them his name they danced in amazement,
touching him to confirm he was not an evil spirit. Lusana
Lumunbala was feasted by the village, and once the cour-
tesies of hospitality were over, was asked what he had discovered in the course of his wanderings and if he had brought
back any treasure. In response he produced a little packet
of dried leaves and a handful of seeds, and addressed the
tribe. They should, he told them, be thankful for the treasure
he had in his hand, which could cure any illness. '"Its smoke,
when inhaled, is to the suffering soul as a mother's caress
is to an ailing child." Saying so, he took a pipe out of his
bag, filled it with a little tobacco, kindled it with some
embers and began to smoke, and as he did so his counte-
nance beamed with happiness.' After smoking, Lusana
Lumunbala explained tobacco's virtue to his tribe: 'This
weed, called Makaya, is man's greatest joy . . . O Makaya,
Makaya, what wonders you can work! As the fire will soften
iron, so Makaya will soften the heart.' The myth concluded
with a moral, coincidentally an exhortation to smoke: 'And
all of you know that Lusana Lumunbala spoke the truth;
whenever your heart rises in wrath or sinks in sorrow, drink
the smoke of Makaya, and peace and happiness will reign
in it again.'

Lusana Lumunbala did not have to explain his pipe, or
the act of smoking to his fellow tribesmen, only tobacco.
Africans already had pipe and smoking cultures of their own.
These revolved around cannabis. Unlike the Europeans, who
did not smoke prior to tobacco's introduction, a number of
African tribes had long had a herbal friend whose fumes they
drank. Cannabis, or *dagga*, as this other combustible weed
was known, was valued for its psychoactive properties, in
particular its ability to generate sensations of well-being. A

dagga smoker enjoyed levels of enlightenment and happiness that those who had never tried the drug could not approach or even comprehend.

Dagga smokers were also responsible for introducing a new weapon into the tobacco smoker's arsenal: the water pipe. Working on the principle that a cool smoke was a good smoke, they incorporated a water chamber in between the pipe's combustion bowl and its mouthpiece, which lowered the temperature of the passing smoke, while simultaneously condensing some of its ingredients. This flavoured residue was then poured on to bark or dried grasses and chewed to generate a second helping of enlightenment.

The figure below shows a Mountain Damara Dagga pipe, made from the horn of a *Bambi*, or antelope, in which a hole has been cut to receive the stem of the bowl. The horn was partially filled with water, into which the stem reached, and the smoker inhaled through the wide open end of the horn. Other tribes used gourds or sections of bamboo as the water chamber, often decorating these with tribal markings or personal insignia. In many cases the pipe itself was no more than an ingenious adaptation of the traditional water vessel a tribe used – converted to a vehicle for spiritual, rather than corporeal refreshment. Once tobacco had become a widely traded commodity the implement devised by cannabis smokers to enhance their pleasure was exported to and imitated in two other continents, metamorphosing into the Persian narghile and the Indian hookah.

This existing smoking tradition assisted tobacco's absorption into African culture. By the 1600s, every tribe the Europeans encountered on the continent were devotees of the weed. Even the bushmen of the Kalahari Desert, who scorned belongings other than their hunting bows and the ostrich eggs in which they transported water, took tobacco, using the Earth itself to smoke – excavating a hole for the bowl, introducing a hollow reed and inhaling prone or on all fours.

Despite tobacco's foreign origin, African pipes were

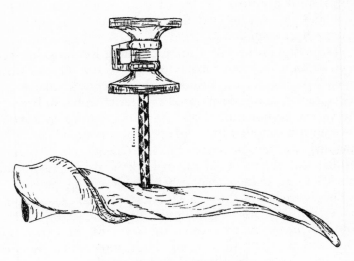

Mountain Damara Dagga pipe

usually domestic in design, and made from materials traditional to each tribe. Africans attached the same importance
to the instrument itself as the North Americans – a pipe
had significance beyond its function. This elevated status
derived partly from the similarity of American and African
social organization – both cultures were tribal and conferred
ritual and ornament on articles of everyday use. In contrast
to the simple clay or shell and tube pipes used by sailors,
who only seemed to care for their contents, the Africans
created highly ornamented devices in a plethora of shapes
and sizes. Some of their pipes were large enough to hold
several pounds of weed, others were equipped with two or
three bowls.

African pipe design reflects in its diversity the plethora
of cultures across the continent. Fetishist tribes made long,
thin, feather hung pipes with bowls in the form of human
heads, Earth mother worshippers made pot-shaped pipes
like pregnant women. Tribal markings were incised into
bowls or carved around stems, identifying the instrument

as an integral part of its owner's culture. The Ashantis, in a manner reminiscent of America's mound-building civilizations, shaped their pipe bowls to resemble totem animals. Vultures, leopards and crocodiles were favourites. A projecting foot like those of America's plains tribesmen was incorporated into the bowls of communal pipes, which could be passed round a circle of seated warriors.

The Africans' penchant for tobacco even shocked the English, whose consumption of the weed was considered amazing by fellow Europeans. A visiting Englishman observed in 1607:

> Tobacco is to them as much as half their livelihood. And the women are as violent smokers as the men. They press the juice out of the leaves when they are green and fresh, then lay them a drying on a shard over the coals, and so cut them for use: for they intimated by signs to us, that the tobacco would make them drunk, if they took it with all its strength remaining in the leaves. The bowls of their pipes are made of clay, and very large; and in the lower end of them they stick a small hollow cane, a foot and a half long, through which they draw the smoke: and this they are not contented to have only in their mouths, but they must have it down to their stomachs too, and so drink tobacco in the strictest sense.

In addition to smoking tobacco, some African tribes developed a snuff habit, including the Masai of Kenya, a tribe of cattle rearers proud of their slim figures and contemptuous of their neighbours. The Masai emphasized their difference by taking snuff, and instead of bestowing their creative skills on pipes, they manufactured beautiful receptacles for tobacco dust.

African enthusiasm for tobacco had tragic consequences. The Europeans' principal African export business was in slaves. The mines and plantations of Spain in the New World had an inexhaustible demand for labour, which the decimated Indians of their colonies could no longer supply. The Portuguese began to ship Africans across the Atlantic, many of whom had been purchased with tobacco. Slaves to a habit, they became real slaves.

4

Bewitched

*The rage of Caliban – King James I of
England attacks smoking – his subjects
leave for the New World to grow tobacco
and to find themselves – the battle against
home-grown – the Dutch invent themselves
and start to smoke – tobacco's debut on
canvas – its persecution in the east by
Christians, emperors and tyrants*

Smoking is a custom loathsome to the eye, hateful to
the nose, harmful to the brain, dangerous to the lungs,
and in the black, stinking fume thereof nearest resembling the horrible Stygian smoke of the pit that is
bottomless.

Thus far, the global dispersal of tobacco had proceeded
smoothly. The weed had been integrated within diverse
cultures, and diagnosed as beneficial by the medical systems
of Europe, of China, and of India. The limited opposition
it had met had been religious. The fact that tobacco had
been overlooked by the founders of Christianity and Islam
troubled the followers of both religions, who allowed their
lives to be governed by the words of their God or His
Prophet, and looked to their spiritual leaders for guidance
on the unexpected. The Koran's omission of tobacco,
combined with Muslims' prior exposure to smoking via their

east African settlements, facilitated its acceptance in Islamic societies. If it was not banned, it therefore must be permitted. Among the Christians, the Roman Catholic Church had taken an equivocal approach to the weed, prohibiting its use in the house of God, but maintaining an official silence on the issue of whether it was a sin or a blessing. Other Christians, however, responded more robustly to the weed. The first to speak out against tobacco was James I, England's Scottish king, who had succeeded Elizabeth I to the English throne in 1603. After making peace with the Spaniards he started a war against the 'deceivable weed'.

As his truce with Spain revealed, James had little appetite for real wars. His preferred realm of conflict was the spiritual. While ruling Scotland as James VI of that country, he had engaged the Antichrist in combat, both in print and through the persecution of his Scottish minions. In 1597, he published a treatise on demonology, that advocated the sternest punishments for Satanists, whom he believed, to the point of paranoia, were determined on his ruin.

James took a personal interest in the discovery of witches, overseeing their torture and editing their confessions. His victims included a Dr Cunningham, implicated in a demonic plot to drown the king while he was on a sea voyage, whose fingernails were pulled out, eyeballs impaled with red hot needles, and who suffered the infamous torture of the 'boots' in which his legs were encased and beaten so 'That the blood and marrow spouted forth in great abundance, whereby they were made unserviceable for ever.' An average of 400 witches per annum were burned in the latter years of James's Scottish reign.

Tobacco was a natural adversary for James. In addition to its visible hell-fire associations when smoked, the tobacco plant had a family tie to witchcraft. Witches' favourite potions for flying were compounded from tobacco's cousins in the *Solanaceae* order – belladona and henbane – whose leaves witches would pound into a paste and smear on broomstick handles, which they then slipped between their legs and rode around their bedrooms, hallucinating flight.

Indeed tobacco was known for a brief period as henbane of Peru. As soon as King James had been crowned in England, he added a law against witchcraft to the English statute books, and opened up another front in his battle against the Devil through the publication of a pamphlet entitled *A Counterblaste to Tobacco*.

The *Counterblaste* is a sour but eloquent composition – 'splenetic' in the language of humours. It commences with an unflattering description of tobacco's pedigree. In the words of a king:

Tobacco being a common herb, which . . . grows almost everywhere, was first found out by some of the barbarous Indians, to be a preservative, an antidote against the Pox, a filthy disease, whereunto these barbarous people are (as all men know) very much subject, what through the uncleanly and adjust constitution of their bodies, and what through the intemperate heat of their climate: so that as from them was first brought into Christendom, that most detestable disease [syphilis] so from them likewise was first brought this use of Tobacco, as a stinking and unsavoury antidote, for so corrupted and execrable a malady, the stinking suffumigation thereof whereof they yet use against that disease, making so one cancer or venom to eat out the other.

Having worked his way through foreigners and syphilis, James's *Counterblaste* arrived at his old enemy, the Devil: 'And now Good Countrymen let us (I pray you) consider, what honour or policy can move us to imitate the barbarous and beastly manners of the wild, God-less and slavish Indians, especially in so vile and stinking a custom? . . . Why do we not as well imitate them in walking naked as they do? In preferring glasses, feathers, and such toys, to gold and precious stones, as they do? Yea, why do we not deny God and adore the Devil, as they do?' In other words, smokers were no better than Devil-worshipping savages.

A universal weakness of humanity is to equate the strength of an argument with their opinion of its maker. Had King James been a popular ruler, the eloquence of his words might have been better received, and he need have continued no further. Englishmen would have thrown away their tobacco pipes and sunk to their knees in penance. Sadly, he was despised, which he knew, and instead of ending his argument, he elaborated.

His *Counterblaste* moved on to wrestle with the concept of addiction. King James felt his way towards this novelty through its predecessor, sin. He succeeded in identifying the essence of addiction – the necessity of the superfluous – and rebuked tobacco users accordingly:

> Thus having, as I trust, sufficiently answered the most principal arguments that are used in defence of this vile custom, it rests only to inform you what sins and vanities you commit in the filthy abuse thereof. First, are you not guilty of sinful and shameful lust? (for lust may be as well in any of the senses as feeling) that although you be troubled with no disease, but in perfect health, yet can you neither be merry at an Ordinary, nor lascivious in the brothels, if you lack Tobacco to provoke your appetite to any of those sorts of recreation . . .?

James compared smokers to alcoholics, who start slowly, but little by little are beguiled and entrapped by their vice: 'And as no man likes strong heady drink the first day . . . but by custom is piece by piece allured . . . So is not this the very case of all the great takers of Tobacco? Which therefore they themselves do attribute to a bewitching quality in it?' There is a sense of frustration running through the *Counterblaste*. Monarchs hate being ignored, especially if they subscribe to the theory, as James did, of the Divine Right of Kings. His strident tone undermines the strength of his arguments, and his pamphlet was ridiculed upon publication. When he realized his rhetoric had failed James resorted to the only other weapon against tobacco at his disposal – taxation. James may have been inspired to tax tobacco by England's physicians, who had sought a monopoly on its supply from him in 1603, asserting that it was a medicine which should only be sold on prescription, and thus giving an indication of the value of the domestic tobacco trade. It is equally possible that James did not need encouragement. He hated tobacco and was an active hater. He also hated sex with women but that was harder to tax. In the same year as his *Counterblaste* was published, he raised the duty on tobacco by 4000 per cent.

The increase in duty made the case for securing an independent tobacco supply, via the establishment of a New World colony, even stronger. Although tobacco was cultivated in England in increasing quantities, most of the country's supply was imported from its continental rivals, constituting a drain on England's finances. When James made peace with Spain, thus removing the risk of reprisals against an English colony, his subjects decided it was time to have another try. The English navy had grown in size whilst the nation had been at war and now vessels were freed up for other purposes. Further, the financial backers of many of England's privately owned warships had to find missions other than piracy to invest in.

Not only were more boats and money available for colonizing projects, but the boats were better and their crews more experienced. By 1603 nearly 100 English boats were making an annual transatlantic voyage to the great cod-fishing grounds off Canada. These fishermen's knowledge of sailing conditions in the Atlantic improved ship design and, collectively, they constituted a pool of experienced sailors whom colonists could use to their advantage. Finally, public opinion in England had changed. The New World was now enticing, instead of uninteresting. It was the subject of plays and essays in which the prospects of not just a new world but a new society too had been examined.

As a result of all these factors, the Virginia Company was incorporated in 1606 with the aim of establishing commercially viable colonies in America, and with the side effect of inculcating the profit motive into its future citizens. In order to avoid the king's censure, the Virginia Company's aims in the New World were to produce wine and olive oil, to weave silk, and to mine whatever metals it came across, precious ones especially. Tobacco was not on this wish list, although it was comprised of articles England presently had to import. The English in the guise of the Virginia Company returned to America for a third time in 1607. This venture consisted of 144 men, of whom 105 reached the New World alive. They founded a settlement in the Chesapeake Bay which they christened Jamestown. Only fifty-three of them lived to see its first anniversary. A relief convoy was sent from England in 1609, composed of nine ships and 500 men under a new governor, Sir Thomas Gates. Gates was shipwrecked in Bermuda, but the remainder of the convoy reached Jamestown and set 400 men ashore. When Gates arrived in the colony six months later with two ships he had constructed in Bermuda, Jamestown was a ruin and its population numbered only sixty men.

It seemed that the new colony was destined to suffer the same fate as the first two. Most colonists died within a year of reaching the New World – from disease, from cold, from starvation and from conflicts with Indians who had quickly tired of having to feed them every winter. The response in England to the colonists' failure to support themselves or even stay alive was one of exasperation. They were exhorted to grow vines and weave silk and the Virginia Company which had sponsored the venture issued regulations against idleness. These included the death penalty for any colonist who missed church more than three times in a row.

However, in 1612, a breakthrough occurred, that was ultimately responsible for the pre-eminence of English culture, language and laws in the most powerful and most imitated nation in history. That year, a Virginian colonist named John Rolfe planted some seeds of *Nicotiana tabacum* that he had obtained from Trinidad. Two years later, he married a teenage Indian princess named Pocahontas ('the frisky one') who had attracted his eye by turning cartwheels naked through Jamestown's streets. Tobacco and love succeeded in accomplishing what sermons and orders had failed to achieve: a self-supporting English colony in America.

Tobacco's potential as a source of profit had been predicted in a visionary statement by Robert Harcourt, whose *Relation of A Voyage to Guiana* anticipated the glories the weed would bring to England: 'I dare presume to say . . . that only this commodity tobacco (so much sought after and desired) will bring as great a benefit and profit to [colonial ventures] as ever the Spaniards gained by the best and richest silver mine in all their Indies.' However, in the hostile atmosphere of King James's court, the exhortation had passed unnoticed.

While other colonists died with their usual abandon around him, or tried their best to find gold to mine or to plant olives as per their instructions from the Virginia Company, Rolfe improved his blending and curing techniques. Tobacco is a difficult crop to raise and to prepare for market. Rolfe took two seasons to learn to grow tobacco and two more years to master the curing process, even when working under the pressures of famine, when the consequences of failure were death.

Rolfe learned some of his planting and curing skills from his Indian in-laws, others from the Spaniards. The Indians recognized tobacco's delicacy and dedicated more effort to its cultivation than for any other crop. Leaves were harvested one at a time and wrapped in bracken fronds for transportation. They were cured by alternate spells in sweat lodges and sunshine. In autumn they were laid outside every morning at dawn to absorb the dew. The seed pods from one year's crop were tied in small bunches and hung in their houses all winter where they became blackened with smoke from the fires. They were taken down, crushed, and the seeds scattered on the planting grounds the following spring. The Spaniards used more sophisticated growing techniques than the Indians – a legacy of their own country's occupation by

a desert nation, which had refined the arts of cultivation
and irrigation. As befits a civilization founded on the concept
of ownership of land, Spaniards were also more orderly in
their planting regimes. Each tobacco seedling was allocated
a precise measure of soil and fertilized with an equal quantity of animal faeces.

Rolfe combined the best of both cultures. He was not just
a visionary, but a scientist. He succeeded in creating a unique
flavour for Virginian tobacco, which quickly gave it precedence in the English market. Some of Rolfe's first tobacco
crop, the 1613 vintage, was shipped to London where its
fragrant odour singled it out for special attention amongst
connoisseurs. In 1616, he took the delectable Pocahontas
and the first commercial shipment of Virginian tobacco to
London. Both products of the New World attracted public
admiration, although the Indian princess succumbed to
disease, died, and was buried in Gravesend. Rolfe returned
alone to America with money and goods in time to celebrate Virginia's first Thanksgiving festival, held to celebrate
the safe harvest of the 1617 tobacco crop.

The same year Captain John Smith, a Jamestown leader,
wrote that the colony's new governor arrived to find 'only
five or six houses, the Church downe, the palisades broken,
the Bridge in pieces, the Well of fresh water spoiled' but, a
lone sign of success, 'the market-place, and streets, and all
other spare places planted with Tobacco'.

John Rolfe's experiment heralded a rapid and permanent
change in the fortunes of England's colonial enterprises.
Englishmen understood the value of tobacco and needed
little persuasion to finance its cultivation. The London
marketplace welcomed increasing shipments of Virginian
weed. The tobacco crop of 1618 was 20,000 pounds. Four
years later, despite an Indian attack that killed nearly one-
third of Virginia's colonists, the settlement sent a crop of
60,000 pounds. By 1627, the shipment totalled 500,000
pounds, and two years later that tripled. 'The discovery that
tobacco could be successfully grown and profitably sold was
the most momentous single fact in the first century of settle-

ment on the Chesapeake Bay . . . Tobacco had guaranteed that the Jamestown experiment would not fail.'

Success with tobacco brought the English colonists new problems. Tobacco planting is extremely labour intensive and required what the colonists lacked most: manpower. A rapid rise in immigration rates was largely offset by an average life expectancy of six months. In 1620, Jamestown's population was 900. By 1624 it was only 1275, despite having received 3500 immigrants in the interim. A typical week was recorded by a colonist in his diary as follows: 'The sixt of August there died John Asbie of the bloudie Fluxe. The ninth day died George Flowre of the swelling. The tenth day died William Bruster Gentleman, of a wound given by the Savages.' The colonists' enemies were not only natural causes. In 1623, the first smoking-related accident was recorded. Seamen celebrating the safe arrival of a Jamestown relief ship in Bermuda lit their pipes in its gunpowder store and 'Drinking tobacco, by negligence of their fire, blew up the ship.'

A solution appeared to Jamestown's labour problem in the form of a Dutch trading ship which dropped anchor in Chesapeake Bay in 1619. The colonists bought twenty 'negars', i.e. African slaves, who were set to work in the tobacco fields. The Dutch traders recognized a promising market and returned in subsequent years with more slaves for sale, and slavery quickly became essential to the colony's economy. In this manner tobacco was responsible for the introduction of slavery to North America.

With slaves to help with the hard work, Jamestown prospered. Life expectancy broke the one year barrier and, for the first time, the colonists had disposable income in the form of tobacco to trade. They spent their earnings wisely and bought women. In 1619, a ship arrived with 'young maids to make wifes' priced at 'one hundred and twenty pounds of the best leaf tobacco'. These bargains were snapped up and Jamestown's church began a birth register. In addition to becoming amorous, the citizens of Jamestown had introduced the notion of independence to America. They created a General Assembly, with legislative powers. Its first law was that tobacco should not be sold for less than 3 shillings per pound. It seems a spirit of invention visited the colonists in 1619. That same year John Rolfe made the greatest ever innovation in the history of tobacco use by introducing the concept of brands.

He named Virginia's product Orinoco, a word, at that time, suffused with the mysteries of Eldorado as described by Sir Walter Ralegh. Lighter in both colour and flavour than its Spanish and Portuguese competitors, it burned with a unique and delicate fragrance, described by a contemporary poet as 'sweeter than the breathe of fairest maid'. Brands *per se* had existed since Roman times, principally for medicines, weapons and wines. However, the concept that identity could improve worth so perfectly fitted tobacco that branding was reinvented by the tobacco trade. Although Rolfe did no more than attach a name to Jamestown's only product, this small step raised it above the level of a commodity. A single evocative word ensured Virginian tobacco was remembered and preferred by the consumer. The English loved Orinoco, or the concept. By 1620 it commanded a premium over every other sort of tobacco.

Despite a loyal market for Orinoco, and the presence of women, Jamestown's situation remained precarious. Deaths exceeded births by a factor of nearly fifty. The Virginians' hardships were compounded by English intervention. King James had not yet learned to love tobacco, but he knew how to profit from it. In 1624 he dissolved the Virginia Company, the commercial operation that had founded Jamestown, and assumed royal jurisdiction. A change of owner meant a new governor for the colonists, and trade restraints. The entire tobacco trade was reorganized as a royal monopoly. King James disguised his intervention as a benign act intended to protect his subjects' interests, while simultaneously acknowledging tobacco to be Jamestown's lifeblood, and therefore safest in his care:

whereas it is agreed on all sides that the Tobacco of these plantations of Virginia . . . which is the only present means of their subsisting, cannot be managed for the good of the Plantations unless it be brought into one hand, whereby the foreign Tobacco may be carefully kept out, and the Tobacco of those plantations may yield a certain and ready price to the owners thereof, That to avoid all differences and contrarieties of opinions . . . we are resolved to take the same into Our Own Hands.

These new restrictions were indicative of the attention being paid in England to its New World colonies. They had become too important a source of revenue to be left to their own devices. This interest in the New World was not confined to England's merchants, or its unpopular monarch. Between 1620 and 1640, 60,000 English men and women left for a fresh start overseas.

Most of these were refugees fleeing King James, his taxes or his son King Charles I who ascended to the English throne in 1625. England was riven by both constitutional and religious discord during the 1620s and 1630s. Many of the English emigrants expected an apocalypse was about to occur and left before the anticipated devastation: 'Judgement is coming upon us . . . This land grows weary of its inhabitants, so as a man, who is the most precious of all creatures, is here more vile and base than the earth they tread upon . . . the fountains of learning and religion are corrupted . . . most children, even the best wits and fairest hopes, are perverted, corrupted and utterly overthrown by the multitude of evil examples.'

Some of these emigrants did not leave England to escape, but to extend the enlightenment offered by Christianity to the 'poor barbarous savages' of the New World. A divine sense of purpose had mysteriously taken root in the English psyche. Englishmen had become convinced that God had chosen them alone to capture souls for Him in the Indies: 'There is no doubt that we of England are this saved People, by the eternal and infallible presence of the Lord predestined to be sent unto these Gentiles in the sea . . . for are not we only set on Mount Zion to give light to the rest of the world? It is only we, therefore, that must be these shining messengers of the Lord, and none but we!'

In 1620, the year after the colonies had received their first Afro-Americans, this unusual type of European settler appeared off Cape Cod in a former claret ship named the *Mayflower*. The ship's cargo comprised 101 men, women and children who, instead of travelling to grow tobacco, or in response to advertising, had left England for the good of their immortal souls. God favoured them in their choice of location. The cape where they had landed enjoyed a mild climate, its water was clean, and the Indians, much reduced in numbers by a smallpox epidemic, were friendly. The Pilgrim Fathers, as these refugees became known, founded the town of New Plymouth, and the countryside around was named New England. Measured by the standards of their age, their settlement was an unprecedented success. By 1630, birth rates exceeded mortality rates. New Plymouth was self-perpetuating. Instead of growing tobacco to trade, it aimed at self-sufficiency, which it achieved almost immediately.

The Pilgrim Fathers' example was followed and by 1630 the English had two models of colony in operation along the American coast. The first type, in Virginia to the south, was based on tobacco, profit and indentured or slave labour. The second kind, in New England, was founded on family values and divine inspiration, and was indifferent, if not actively opposed, to tobacco. Massachusetts, a northern settlement, banned public smoking in 1632, followed by its neighbour, Connecticut, which allowed its citizens to smoke no more than once a day and then 'not in company with any other'.

In addition to settling the mainland, the English founded a series of colonies on Caribbean islands. These were commercial tobacco growing enterprises funded by the country's merchants and sponsored by its peers. The first was on St Christopher, known as St Kitts, a striking island composed of lowlands to either side of an attractive volcanic cone. St Kitts was settled in 1623. Barbados, a solitary island whose rain forest terminated in sandy beaches, was colonized shortly afterwards. Both of these settlements were financed by the City of London. They attracted over 30,000 emigrants during the 1630s, principally in the form of indentured labour for tobacco plantations.

The Caribbean colonies functioned along similar lines to Virginia, with some interesting local complications. St Kitts was a favourite haunt of the Caribs, a maritime race of cannibals, who saw the pale-skinned arrivals as an opportunity to improve their diet. They had eaten the French settlers of a nearby island and had grown partial to white flesh. The Caribs were also fond of tobacco which they carried off from the colonists' plantations in daring nocturnal raids. Barbados, in contrast, 50 miles further out in the Atlantic, had been uninhabited since its creation. Barbados was carpeted with virgin forest which took time to clear for tobacco fields and, owing to the island's small size, could not be simply burned off without taking the colonists with it.

These island settlements were distinctly more feudal than those in either Virginia or New England. Feudalism, with pastoral overtones, was a third system of government attempted by the English in the New World. This multiplicity of colonial models was uniquely English. Columbus's discovery had stimulated an immense body of literature and thought in England, commencing with Sir Thomas More's *Utopia*, a political fable that purported to describe the perfect society, based on slavery. *Utopia*'s idealism was counterbalanced by *Arcadia*, the most popular novel of the period. *Arcadia* described a rustic idyll in which two princes visiting a new country disguised themselves as shepherds in order to complicate their love lives. *Arcadia*'s heroes, however, did not limit themselves to passive admiration of their unspoilt surroundings. Their virgin territory contained virginal princesses whose pursuit and violation became the princes' mission. Even Shakespeare had resorted to the New World as the setting for his final play, *The Tempest*, whose plot was based on the shipwreck of Sir Thomas Gates in Bermuda – which had since been colonized and planted with tobacco. The consequence to English minds of texts like *Utopia*, *Arcadia* and *The Tempest* was an optimistic view of the New World and its suitability to their individual constitutional fantasies.

A feudal venture was also started on the mainland in the colony of Maryland. Founded in 1634 by Lord Baltimore, whose personal vision of America was as a refuge for English Catholics, Maryland was something of a utopian project, which aimed to create a New World that was already at least a century out of date. Lord Baltimore's proposal for Maryland – a hierarchical agrarian society, a copy of the rural English life represented in contemporary fiction as the perfect social template – was a regressive step. Feudalism was being displaced all over Europe. It is hard to imagine life under a feudal system, except as a prince, but relatively easy to see why people wanted to leave one. No one wanted to be peasants for ever in the New World. Lord Baltimore's feudal design failed and Maryland grew into a community of largely Protestant tobacco farmers.

Feudalism, however, was alive and well in England, where King James continued to pursue his vendetta against tobacco. Demand amongst his subjects for the weed had become so great that they were cultivating it commercially within his English realm. By 1630, great parts of Gloucestershire, Wiltshire and Worcestershire were given over to tobacco plantations. While the quality of English weed was inferior to Virginian, it had one significant advantage to the user: the only tobacco taxes in place were import duties, thus rendering home-grown, in effect, duty free.

The practice alarmed England's government. Its monarch expected his subjects to pay for their pleasures and every pipe of home-grown smoked resulted in a loss of revenue. Faced with the option of imposing a tax on tobacco growing, or banning it all together, the government elected for prohibition. James I, who must have felt that tobacco had only appeared in the British Isles to vex him, issued proclamations in 1620, 1621 and from his deathbed in 1624 forbidding the domestic production of tobacco. At the same time he legislated against tobacco smuggling, for the weed flooded into England over every unguarded inch of coastline.

James's subjects took as much notice of his proclamations as they had of his *Counterblaste*. His son King Charles I was forced to issue a similar prohibition in 1633. Its language shows he shared his father's dislike of the deceivable weed, in particular its propensity to provoke men and women to lust:

> The plant or drug called tobacco scarce known to this nation in former times, was in this age first brought into this realm in small quantity, as medicine, and so used . . . but in the process of time, to satisfy the inordinate appetites of men and women it hath been brought in great quantity, and taken for wantonness and excess, provoking them to drinking and other incontinence, to the great impairing of their healths and depraving them of their manners, so that the care which His Majesty hath of his people hath enforced him to think of some means for preventing of the evil consequences of this immoderate use thereof.

Needless to say the best way Charles could think of protecting the health of his wanton subjects was by ensuring they could only smoke tobacco that had been properly taxed. His soldiers continued to burn tobacco fields, and hostilities against the plant were only suspended by the onset of an English civil war.

Although no other European monarch had imitated James I of England in issuing counterblasts, tobacco taxes, and restrictions on the tobacco trade, were being imposed nonetheless across Europe. In Spain, King Philip III decreed in 1606 that tobacco could only be grown in specific colonies, including Cuba, Santo Domingo, Venezuela and

Puerto Rico. Sale of tobacco or tobacco seed to foreigners
was punishable by death. Incidentally, John Rolfe had
obtained his original seed in contravention of this prohibition. A few years later, the Spanish crown imposed a special
tax on tobacco in addition to the *Almojarifazgo* duty that
every product from the New World carried. The rate varied
depending upon the tobacco's place of origin, fluctuating
between 7½ per cent for Cuban tobacco and 2 per cent on
weed from Santo Domingo. Shortly afterwards, further
restrictions were added: all tobacco had to be imported into
Spain through Seville. In 1620, the first Sevillian tobacco
factory was opened in a former penitentiary for fallen
women. The San Pedro factory produced snuff for domestic
sale and for export, and initially employed only men. The
use of snuff had spread from clergy in the New World to
their counterparts in Seville, whose excesses with powdered
weed provoked outrage in the Holy See. Only when Pope
Urban VIII threatened to excommunicate those who smoked
or took snuff in sacred places, on the grounds that these
habits provoked sexual ecstasy, did Sevillian priests desist
from snuffing while saying Mass.

Once it had assumed control of tobacco production, the
Spanish monarchy set out to corner the retail trade. In 1636,
it founded the Tabacalera, the world's first tobacco company,
and introduced state tobacco shops named *estancos*, where
the Tabacalera's products were sold and an additional tax
of 3 reals per pound was collected. Spain's example was
followed in Venice, which introduced a tobacco tariff in
1626, and in France, where the infamous Cardinal Richelieu
imposed customs duty on tobacco in 1629. Within decades,
most European countries in which tobacco was established
had introduced some form of taxation on its production,
import or consumption. European governments were
becoming addicted to the weed as surely as their citizens.
They, however, did not struggle to find justifications for
tobacco. It provided a predictable and increasing stream of
income, which was mainly applied to financing war. Much
of the continent was involved between 1618 and 1648 in a

conflict named the Thirty Years War, which had been provoked by a combination of religious differences and imperialism. The Thirty Years War saw tobacco use spread to the European countries it had not yet reached, such as Sweden, and the general advance of the smoking habit.

Smoking as a method of tobacco consumption was spread principally by the armies of a new nation of tabagophiles – the Dutch. Their country, the seven United Provinces of the Netherlands, was a new nation, carved in 1581 out of a part of Europe that had previously been known by the depressing collective of 'the Lowlands'. The Dutch were proof of the theory that it is not beauty of surroundings that make for quality of life. A slave may be kept in a palace, the Dutch by contrast lived in a swamp bordered by a shallow and turbulent sea, but, for their time, they had more individual liberty than any other nation on earth. Holland was organized on a different basis to the monarchies that governed the rest of Europe. Its rulers were elected and its riches came from trade.

The Dutch were Protestants, and were on friendly terms with their fellow Protestants, the English. In each of these nations education had been removed from the control of religious orders and at their universities both freedom of thought and smoking flourished. An exchange of students between England and Holland stimulated the Dutch tobacco habit and a growing presence in overseas trade furnished them with the supply. It is debatable which of the two countries elevated the medical reputation of tobacco to greater heights, but the Dutch certainly led the field in paediatrics and infant pipe smokers were counted as one of the curiosities of seventeeth-century Amsterdam.

Having received the weed from English hands, the Dutch knew nothing of the rituals surrounding its original use. It was introduced to them as a substance for pleasure alone, and received an enthusiastic welcome. Smoking spread through Holland with the speed of a forest fire and soon the Dutch too were astonishing visiting Germans: 'I cannot refrain from a few words of protest against the astounding fashion lately introduced from America,' wrote Rusdorff, Palatine Ambassador to the Hague, 'a sort of smoke tippling, one might call it, which enslaves its victims more completely than any other form of intoxication, old or new. These madmen will swallow or inhale with incredible eagerness the smoke of a plant they call Herba Nicotina, or tobacco.'

Tobacco formed a Protestant bond, and the act of smoking assisted the Dutch in establishing an identity for their nation. They favoured the pipe as a tobacco delivery system, in imitation of the English, and soon its use was universal throughout the Lowlands. 'A Dutchman without a pipe is a national impossibility. If a Dutchman were deprived of his pipe and tobacco, he would not even enter paradise with a glad heart.'

Just as tobacco and smoking had made their debuts on the English stage, the weed surfaced in Dutch painting, co-incidentally as this was commencing its golden age. The secular spirit that characterized the Dutch state also animated its artists during the seventeenth century. While painting in other European countries had lapsed into mannerism, Dutch artists drew inspiration from representing the common man indulging in his favourite pastimes, i.e. drinking and smoking. Their insistence in placing pipes in the mouth of most of their subjects did much to perpetuate the reputation of the Dutch as smokers *par excellence*.

This culture of the common man was a radical innovation in European art, which had hitherto restricted its represen-tations of real individuals to the wealthy, powerful or pious. For the first time, it was acceptable for painters to choose ugly subjects and they revelled in this new found freedom. David Teniers the Younger's peasants in the figure below are so rustic that they appear only recently to have been dug out of the soil. It is symbolic of the cultural distance tobacco had travelled that its first representation in European art was not, as it had been with the Mayans, in the company of gods or nobles but between the lips of drunken peasants. These charming paintings are the first depictions of tobacco in its modern role: an indispensable pleasure of the masses.

Smoking had become a casual act of stimulation, whose ubiquity was accepted without question. Perhaps this was only the spirit of the age. 'O reason not the need,' cried Shakespeare's King Lear, 'man's basest beggars are in their poorest things superfluous.'

Holland had been born in troubled times. In addition to spending much of its infancy at war, it was afflicted by the bubonic plague, which resulted in a further stimulus to tobacco use. When the plague visited Holland, Dutch physicians put their faith in tobacco, and, if their accounts are to be believed, it was rewarded. Isbard von Diemerbroek, writing as the Black Death raged at Nijmegen in 1636, declared:

> As I have proved by long experience, tobacco is the most effective means of avoiding the plague, providing the leaf is in good condition . . . One day, when I was visiting one of [its] victims, the reek of the pestilence seemed to overpower me, and I felt all the symptoms

Smoking as the favourite amusement of the Dutch

of infection – dizziness, nausea, fear: I cut my visit short and hurried home, where I smoked six or seven pipes of tobacco. I was myself once more, and able to go out again the same day.

Rampant Dutch demand for tobacco necessitated a steady supply. However, instead of imitating the English and starting New World colonies, the Dutch relied on obtaining their tobacco through capture and trade. In 1621, after the expiry of a twelve years' truce with Spain, the Dutch attacked their former masters overseas. They founded a West India Company, later to serve as a model for the English and French in their ventures abroad. The West India Company's aims were plunder, conquest and trade. It was a successful start-up, and after a series of campaigns, one of its senior executives, Admiral Pieet Heyn, supervized the capture of the entire Spanish treasure fleet, an enormous prize whose value was greater than that assigned to the Lowlands a decade previously. This onslaught did much to protect the colonies of other nations from Spanish attention, as the Dutch virtually cleared Spanish shipping from the Caribbean. In addition to providing a protective umbrella under which the English and French island settlements flourished, the Dutch alleviated the perennial labour shortage in the New World. The third corporate strategy of the Dutch West India Company was slaving. After it had displaced the Portuguese in Africa, the West India Company began shipping slaves to the Americas in volume. These slaves were bought with tobacco and sold in Virginia for Orinoco.

The Dutch were pragmatic in their view towards the New World. Instead of attempting romantic missions or constitutional experiments, they acted as accomplices in its despoliation. They encouraged foreign colonists to stake their lives on crops, and, if these later succeeded, bought them in cash for a ready market. It seems that life in Holland was too pleasant to abandon for a brutish existence in a distant continent. The Dutch established a few outposts in the New World – Curaçao, Saba, St Maarten and St Eustatius in the

Caribbean, and New Holland on Manhattan Island in North America. All of these were little more than trading posts, although New Holland supported a few tobacco farms. The original name of Greenwich Village was *Sapponckanican* – 'tobacco fields' or 'land where the tobacco grows'. The Bowery area of New York City was also home to a tobacco farm, founded by the Dutch governor Wouter Van Twiller in 1629.

Tobacco thrived in Europe and the Americas throughout the first half of the seventeenth century, not only as a consequence of the efforts of Virginian colonists and Dutch traders. Governments had acquired an interest through taxation in popularizing the plant, which was also dispersed by war, and by sporadic plague epidemics. However, in other countries and on other continents, the spirit of King James I, in the senses of hatred and oppression, led to extreme countermeasures against tobacco. James's fellow Christians, in Russia for example, suffered both temporal and spiritual persecution.

Michael Feodorovich, the first Romanov tsar (1613–45), had no interest in taxing the weed, and persecuted tobacco users instead. Its consumption, in any form, was declared a capital offence. A Tobacco Court was established to try breaches of the law. Its favourite punishments were slitting the lips of a smoker, or an excruciating, and usually fatal, flogging with the knout. Occasionally, offenders were castrated, but if they were rich, they were exiled to Siberia and their property was confiscated. Russians who dared to smoke in breach of their tsar's prohibition faced the incidental risk of damnation. The Greek Orthodox Church, to which all Russian Christians belonged, had coincidentally decided, after scrutinizing the Bible, that it was tobacco smoke which had intoxicated Noah, causing him to reveal his genitals to his children, and therefore had banned its use in 1624.

Smokers also suffered in Islamic countries, where, in the hands of the right despot, the habit's other-worldliness, in

the sense of its absence from the Koran, invited elaborate
punishments. If it was not permitted, then it must be
forbidden. The most extreme persecution of early smokers
took place in the Ottoman Empire, an Islamic state that
stretched from Jerusalem to Constantinople. Murad IV, also
known as 'Murad the Cruel', ruler of this vast territory
between 1623 and 1640, had despised smoking ever since
the firework display celebrating the birth of his first son had
burned down half of his capital, Constantinople. Murad's
habit was to roam the streets of Constantinople in disguise,
feigning an urgent craving for a smoke then beheading any
good Samaritans who offered relief. Contemporary records
state that he himself killed or had put to death, in the space
of only fourteen years, in excess of 25,000 suspected
smokers. Murad's example was followed in Persia, where
merchants caught selling tobacco were executed by having
molten lead poured down their throats.

Tobacco also ran into trouble in the Far East. Japan's
rulers were becoming hostile to foreign influences and
tobacco's alien origin counted against it. The weed was
banned in Japan in 1609, 1612, 1615 (twice) and 1616,
each time with increasingly severe penalties. By 1616, the
punishment for smoking was a fine, imprisonment and
confiscation of all of the smoker's remaining possessions.
The prohibitions were ignored. By the time of the fifth ban
the Shogun's own bodyguards were smoking openly in front
of him, an amazing liberty in a court that punished even
suspected insolence with decapitation. In 1625, the ban was
repealed and within fifteen years the Japanese had adopted
the Native American custom of offering visitors a pipe of
welcome.

Tobacco's fortunes suffered similar fluctuations in China,
where the people's love affair with the weed had been interrupted by the introduction of the death sentence for smoking
by the last Ming emperor in 1640. However, their passion
for tobacco could not be frustrated: 'If we shut Nature out
at the door, she will come in through the window.' The
emperor's subjects investigated other methods of tobacco

use, and took to snuff as a substitute for smoking. The jade, amber and rock crystal receptacles they created to contain this precious powder represent the appearance of a new form of art. Happily for the Chinese, the penalty of decapitation for smokers did not last. In 1644, the Ming Dynasty was succeeded by Qing rulers who were Manchus by blood, and who had developed a taste for tobacco in their native Siberia. After conquering China they legalized smoking and encouraged tobacco cultivation. Their favourite strain of the weed, developed in Fukien Province, was the famous Silk Cut, a pale, fragrant tobacco which had been shredded so finely that it resembled fibres of silk, and which was used by courtesans to while away the idle hours they spent awaiting their Manchu lovers.

5

Middlemen and Foreign Bottoms

*Tobacco as an instrument of trade
between all nations – England's love
affair with the weed continues – the
consequences for its neighbours, in
particular, the Scots and Welsh – tobacco
as an item of plunder – tobacco as currency*

The disruption to shipping resulting from the English civil
war enabled the Dutch to build their trading links with its
colonies. Ties between the English colonists and the Dutch
had begun with Jamestown's slave purchase in 1619 and
had grown steadily since. The Dutch provided what the
motherland could not, their goods were cheap, and they
were willing to accept tobacco in exchange for slaves, who
had become necessary to planters as the supply of indentured Englishmen dried up in the civil war years.

The Dutch recognized tobacco's value as an instrument
of exchange. It was a product which had a value wherever
they carried it, and they made it the first globally available
luxury. They not only bought tobacco to satisfy their own
market, but as an item of trade for use in the rest of their
mercantile empire, which by 1660 was the most comprehensive in the world. They had taken over most of Portugal's
trading stations in Africa and Asia, whilst adding to these
by installing new outposts of their own. The Dutch dealt
with Africans, Arabs, Indians, Ceylonese, Indonesians,

Chinese, and had a monopoly on European trade with Japan. In some of the countries with which they traded, tobacco was the only acceptable commodity the Dutch possessed. Jan Van Riebeck, an executive of the West India Company observed in 1652: 'If we had no tobacco there would hardly be any trade, as a whole cow would often be withheld for a finger's length of tobacco or a pipe'. Tobacco could be exchanged for food, ivory, silk, spices and slaves at a very profitable rate: 'today we bought a fine sturdy milch ewe for its length in twisted tobacco, not more than ¼lb in weight and only worth 11 doits'.

Tobacco enabled the Dutch to make some amazing bargains, in Africa in particular. In 1652 the Dutch purchased the entire peninsula of the Cape of Good Hope for 'a certain quantity of tobacco and brandy'. The Hottentots, the nation with which they had struck this bargain, were a remarkable race who had integrated tobacco into their unusual culture in a manner which has never been imitated. On achieving puberty, a Hottentot boy was given his first cigar while his mother bit off and ate his left testicle. As Guy Tachard, a French Jesuit, observed: 'they have some very odd and whimsical customs . . . The men in their youth make themselves half eunuchs, pretending that that contributes much to the preservation and increase of bodily agility.'

Tobacco was one of the few things that would persuade a Hottentot to indulge in manual labour:

> The Hottentots, being persuaded that there is no life after this, labour as little and take as much ease as they can in this world. To hear them talk even when they are serving the Dutch, for a little bread, Tobacco or brandy, they look upon [the Dutch] as slaves who work the land of their country, and as people of no courage who shut themselves up in houses . . . whilst their people encamp securely in the open fields without stooping so low as to labour land. By that way of living they pretend to demonstrate that they are Masters of the Earth and the happiest people in the world.

The Hottentots had a sad demise. They fell in love with the white man's vices and drank, whored and smoked themselves into monotesticular oblivion.

The Dutch found tobacco equally useful as an item of commerce in Asia, where they had established new settlements, including Batavia in Indonesia – a fine and spacious city separated by a ditch from a slum named Jakarta. Batavia formed the capital of the Dutch East Indies and from here tobacco use was spread to those areas which had not received the weed from the Spanish, Portuguese or local trade. Dutch ingenuity in creating markets for tobacco where none had existed previously was remarkable. The Balinese were persuaded to chew tobacco after taking betel-nut on the assurance that the weed cleaned their teeth. The Javanese, who were addicted to cloves, were provided with tobacco mixed with pieces of clove. These fragments made an attractive crackling noise upon combustion, which was believed to ward off evil spirits and bring the smoker good fortune.

Initially, the Dutch encouraged tobacco's cultivation in the Far East. As early as 1625 they were shipping tobacco grown in their settlement at Jaffna, in Sri Lanka, back to Europe. The Sri Lankans were as enthusiastic smokers as their Dutch masters. Their preferred method of consumption was a slender form of cigar known as the cheroot, whose name derived from the Tamil word *shuruttu*, meaning 'to roll'. They were described indulging in their favourite pastime as follows: 'The poor sort of inhabitants, viz. the Hindoos, Malabars etc smoke their tobacco after a very mean but I judge original manner: only the leaf rolled up and light one end, holding the other between their lips.' Later, however, this home-grown weed began to compete with Dutch trade tobacco and the Dutch prohibited its transportation on their ships.

The Dutch became masters of the Tropic of Cancer, within which they distributed tobacco and other trade goods in exchange for the treasures of the Orient. They were conscious of tobacco's compelling power on its users, which they sometimes used to unlock new territories. Once tobacco had been introduced, repeat orders were a certainty, and there was very little it could not buy. Tobacco was also universally acceptable as a gift – a useful attribute in an age when dietary differences caused wars. Even ornaments could be dangerous territory, if they inadvertently trivialized some sacred item of the local culture, such as a cow. But a tobacco pipe, or even plain tobacco, made a welcome present almost anywhere the Dutch visited.

Tobacco did not only serve as an item of commerce between Europeans and other races. Several Asian countries grew tobacco to trade as well as for internal consumption. It was transported across the Indian ocean in Arab dhows, and from country to country and island to island through Malaya, Java and Borneo to the northern tip of Australia. Asia was the second link in the chain of tobacco use. Asian countries were introduced to it by their neighbours, and as a consequence, several nations got their first taste of the weed third hand. Both the reasons for and procedures of smoking adopted in parts of Asia were the result of a mixture of American, European and Chinese superstitions.

Although the Europeans had introduced tobacco to many Asian nations without any justifications for its use beyond pleasure, it was transferred between Asians with reasons for its consumption attached. The Chinese told the Vietnamese that tobacco was good for their health and dyed it red, the colour of success, before shipping it to Saigon for sale. They also marketed it in Burma as a medicine, where its use became commonplace amongst both sexes. Burmese women nourished their children with alternate sucks from teat and cheroot, in the belief that a peaceful infancy enabled a child to develop strength.

As tobacco travelled it also acquired new properties and legends were attributed to or grew up around the weed. The Malays believed tobacco was created in China by the fortunate union of a dragon and a snake. Torres Islanders pointed due north when asked to name its place of origin. The variety of races that might share the same tobacco crop was astonishing, and the weed came to be associated with friendship and peace. Tobacco spread like an epidemic in the seventeenth century, crossing borders and cultures at will. Like a contagious disease, it respected neither rank nor creed, infecting nomads, Buddhists and Christian bishops alike. By the middle of the seventeenth century, smoking was becoming a defining human habit.

In addition to being dispersed via maritime trade, tobacco became a staple of the silk route and travelled overland in merchants' caravans and camel trains, reaching places which had never been seen by Western eyes. Tobacco became the confidante of the harem and audience chamber and was present at the councils of a hundred rulers. Some countries received tobacco from both the west and the east, on each occasion with different explanations of its origin and purpose. In Persia for example, tobacco had first arrived from the west, and both smokers and tobacco merchants had been subjected to draconian punishments meted out by the Shah. However, when re-introduced from the east, in combination with a water pipe adopted from the Moghuls of India, tobacco smoking became a favourite pastime of the languid Persians. An English visitor, Dr John Fryer, was impressed by their smoking dens:

> modelled after the nature of our theatres, that everyone may sit around and suck choice Tobacco out of long Malabar canes fastened to crystal bottles . . . where the bowl or head of the pipe is inserted, a shorter cane reaches to the bottom, where the long pipe meets it, the vessel being filled with water. After this sort they are mightily pleased; for putting fragrant and delightful flowers into the water, upon every attempt to draw tobacco, the water bubbles, and makes them dance in various figures, which both qualifies the heat of the smoke, and creates together a pretty sight.

Tobacco was reintroduced to the Ottoman Empire from Persia. After the death from alcoholism of Murad the Cruel in 1640 and a brief battle of succession, the Ottoman throne was assumed by Mohamed IV, a dedicated smoker, who encouraged tobacco's cultivation and scientific experiments to improve the curing process. The Ottomans adopted the Persian water pipe and renamed it 'narghile'. As a backlash against the danger that had surrounded smoking under Murad IV, the narghile was a pipe of peace, whose users dedicated themselves to relaxation and to achieving the perfect trance. As a contemporary commentator noted, tobacco had joined coffee, wine and opium as one of the four 'cushions on the divan of pleasure'.

Manufacture of the narghile evolved into an art form, or rather, four art forms as the pipes' constituent parts were manufactured in different parts of the Ottoman Empire by specialized craftsmen. The *agizlik*, or mouthpiece, was carved from amber imported from Russia. The *govde*, or

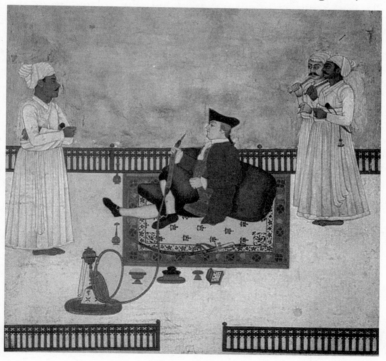

Dream weaver

water bowl, was hand blown in glass or crystal and chased with flower motives. The best *govdes* were produced in Beykoz, at the opposite end of the empire to Tophane, where the *lule* (pipe bowl) makers resided. The final constituent part of the narghile, the *marpuc*, a woven tube which delivered the smoke to the user, was produced by nomadic tribesmen who traded their creations in the souks for henna.

Once the constituent parts of a narghile had been created and assembled, they were delivered to the restful saloons patronized by Ottoman smokers. In these dark and fragrant sanctuaries, a complex etiquette was devised which extended to the correct manner in which the pipe should be lit, even specifying the type of wood burned to provide the charcoal required for this function. Narghile smokers developed a preference for the dark strong tobacco grown in Persia known as 'latakia', thus creating a market entirely isolated from the weed's place of origin, which was forgotten, and the pleasure-loving Ottomans would have been astonished to learn that tobacco had been discovered by the Incas, a distant and extinct civilization, instead of being a traditional crop of the Levant.

Following its reintroduction to the Ottoman Empire, tobacco continued to travel westwards. Russia had no sooner banned the weed in its Baltic provinces when it became the staple trade good of any camel train from the east. Fortunately for tabagophiles, when Peter the Great became Tsar of Russia in 1689, he repealed his predecessors' statutes imposing dreadful punishments on smokers. This new freedom to smoke was opposed by the Russian Orthodox Church, which persisted in the belief that smoking was specifically prohibited by the Bible, and had discovered a condemnation of tobacco in the New Testament as well as the Old. The key text was to be found in the Gospel of Mark, 6:15: 'The things that come out of him, those are they that defile the man.' The Patriarch of Moscow threatened smokers with excommunication, thus provoking a trial of strength with the tsar. But Peter did not enjoy, or even allow, disagreement. He himself smoked, having been introduced to tobacco by a Scotsman, General Patrick Gordon of Auchterleuchries. Peter responded to the patriarch's challenge by imposing a tax on beards of the sort favoured by the Orthodox clergy. If he observed a beard worn in his court he himself would trim it with a pair of scissors he carried specially for that purpose.

Arrival by land, however, did not guarantee tobacco a welcome. Its reception in Switzerland was hostile, and when the Swiss found the weed hard to resist, smoking was prohibited, and the ban was enforced by a dedicated tobacco court. Swiss opposition had been inspired by the German state of Saxony, where, in 1653, German Catholic priests decided that tobacco was no friend of mankind, voicing their objections on the twin grounds of piety and sanitation: 'it is both godless and unseemly that the mouth of man, which is the means of entrance and exit of the immortal soul, that mouth which is intended to breathe in the fresh air and to utter the praises of the most high, should be defiled by the indrawing and expelling of tobacco smoke'. Their pronouncement was reinforced with prohibitions of varying severity. Smoking even carried the death penalty in Lüneburg, a law that survived until 1691.

In general, nevertheless, if European governments had any official view on tobacco, it was biased towards its potential as a source of income. Why allow religious sentiment to interfere with such a lucrative substance? Some European countries, including Austria, whose capital, Vienna, was besieged by the now smoking Ottomans in 1683, actually encouraged the nascent smoking habit. Emperor Leopold I introduced a national monopoly on tobacco's sale, with the aim of ensuring it was as widely available as possible throughout his realm. The money he raised from selling this monopoly was applied to financing his hunting expeditions. Leopold's pretext for profiting from tobacco was that the plant was a luxury:

> We have graciously determined, in virtue of Our Imperial, Royal and Princely power and sovereignty . . . to draw a revenue from tobacco, whether smoked

or taken as snuff. This commodity is not needful for the sustenance of mankind – rather it has become an arbitrary though almost universal habit. For this purpose We have determined, through our Imperial Exchequer, either Ourselves to take over the trade in such tobacco, or, according as We may think most profitable to Our treasury, to lease or farm it out to one or more persons.

Leopold of Austria's assertion that tobacco was not 'needful for the sustenance of mankind' would have raised eyebrows in England, where affection for tobacco appears to have been rejuvenated by civil war to the extent that the weed was no longer considered simply a medicine or intoxicant, but an absolute prerequisite of health. Despite a century without proof of its efficacy, the English had elevated tobacco to the status of panacea. In keeping with such expectations, tobacco was also believed to enhance longevity. A picture of the oldest recorded man in England, Thomas Parr, reputed to have lived from 1483 to 1635 and who 'did penance when he was a centenarian for begetting a bastard child', was a popular sign for tobacco shops.

Curiously, the English did not make any connection between smoking and the recent discovery of the purpose of human lungs, which had been made by William Harvey as a part of his explanation of the function and circulation of blood. Hitherto, the best analysis of why humans had lungs had resided in the textbooks of Chinese medicine, which held them to be the residence of a mysterious and vital force named 'chi'. However, instead of investigating the relative merits of inhaling air or smoke, the English put their trust in new proofs of tobacco's infallibility. Following the publication in English of Dutch testimonials, tobacco was accepted as being especially effective against bubonic plague, which killed a sixth of London's population in 1665. Even Samuel Pepys resorted to his countrymen's favourite precaution in the face of an epidemic:

This day, much against my will, I did in Drury Lane
see two or three houses marked with a red cross upon
the doors and 'Lord have mercy upon us' writ there;
which was a sad sight to me, being the first of the kind
that, to my remembrance, I ever saw. It put me into
an ill conception of myself and of my smell, so that I
was forced to buy some roll tobacco to smell and to
chaw, which took away the apprehension.

With or without a plague epidemic, the English were so
committed to smoking that they once again were amazing
visitors from the continent:

> in England, when the children went to school, they
> carried in their satchels, with their books, a pipe of
> tobacco, which their mothers took care to fill early in
> the morning, it serving them instead of a breakfast; and
> that at the accustomed hour every one laid aside his
> book to light his pipe, the master smoking with them,
> and teaching them how to hold their pipes and draw
> in the tobacco; thus accustoming them to it from their
> youth, believing it absolutely necessary for a man's
> health.

Englishmen even kept pipes beside their bed, which they
used to punctuate a good night's sleep.

Tobacco's influence over English learning extended
beyond the classroom. It reacquired an association it had
enjoyed among the Aztecs, whereby smoking was identified with meditation. As, to English eyes, smoke was drunk,
it must therefore nourish something, but as it had no
substance, this therefore only could be the spirit. Isaac
Newton, the greatest scientist since Aristotle, smoked inces-
santly. Whether this habit contributed or not to his inven-
tive genius is impossible to determine, for Newton smoked
from his infancy until his death bed. The weed was so integral to his identity that he was once observed 'in a fit of
mental abstraction, using the finger of the lady he was

courting as a tobacco stopper, as he sat and smoked in silence beside her'.

Tobacco continued its illustrious association with England's poets and playwrights throughout the seventeenth century. It was acknowledged as an inspiration by, among others, Robert Herrick and John Milton, respectively the greatest lyric poet and the greatest poet of the age. Herrick celebrated tobacco's ephemeral qualities:

> It is all spirit, not to force belief,
> It is the life of air, the air of life.

Smoking became a metaphor for mutability – the evanescence of pleasure, indeed, of existence itself – and was a favourite subject of meditations upon transience:

> Here, all alone, I by myself have took,
> An emblem of myself, a Pipe of Smoke.
> For, I am but a little piece of clay
> Filled with smoke that quickly fumes away.

The English were unable to imagine a life without tobacco. In order to make Robinson Crusoe, the story of a castaway on a desert island plausible, Daniel Defoe allowed his hero to discover 'a great deal of tobacco, green and growing to a great and strong stalk'. The first thing Crusoe does in his isolation is make a clay pipe. Inspired by his first smoke, Crusoe is empowered to make the best of being marooned and builds a goatskin parasol, a boat, a fort, befriends a savage, wages war, and displays all the superior vigour expected of the white man.

Resurgent demand for tobacco required England to repair its post-civil war trading links with the New World. Before this was achieved, other methods of tobacco supply were revived. Some Englishmen had enjoyed the taste of anarchy the civil war had offered and continued to cultivate and smuggle tobacco during Oliver Cromwell's protectorate, and after the restoration of the monarchy. Tobacco was planted throughout the west of the country, provoking a predictable official response to the threat of untaxed tobacco. In 1667, King Charles II was forced to send his Household Cavalry

> down to Winchcombe to spoil the tobacco there which it seems the people there do plant contrary to the law, and have always done, and still been under force and danger of having it spoiled, as it hath been often times, and yet they will continue to plant it.

Tobacco smuggling was also a problem for the English government, which it turned to its advantage by using as a justification for renewing its grip on colonial trade. In order to counter smuggling, especially of tobacco, a Navigation Act was introduced in 1651. The Act decreed that imports into England were only to be brought from the country in which they were produced and carried only in English ships, or ships of the country of their origin. The Act specifically prohibited the use of 'foreign bottoms', by which it meant ships of intermediary nations, in particular those of the Dutch. The Navigation Act also decreed that all goods imported from Africa, Asia or the American colonies could only be carried by English ships.

The Act had immediate repercussions, the first of which was war with Holland, which had been England's faithful ally for eighty years and had, through its attacks on the Spaniards, ensured the survival of England's New World colonies. The Act likewise inconvenienced the colonists who had come to rely on the Dutch for most of their necessities in the civil war years and who could not be serviced by the much smaller English merchant fleet. While Englishmen had been killing one another at home, the colonists had enjoyed a degree of autonomy and had found this taste of freedom, or anarchy, to be as addictive as their principal export.

A second Navigation Act was passed in 1660 which imposed even further restraints on the colonies' trade, by prohibiting the settlers from selling their tobacco to any other nation than England. The enterprising colonists had meanwhile developed new markets for their Orinoco, in particular in continental Europe where Virginian tobacco had acquired a reputation for consistent excellence. They therefore especially resented the additional restrictions the second Act had imposed, as it reduced their status to that of crown servants, fit only for carting dung and raising tobacco plants.

The two Navigation Acts made the English three enemies: Holland, their colonies and Scotland. Scottish merchants had taken up some of the slack in the tobacco trade during the civil war years and suffered accordingly when their country was deemed a foreign nation for the purposes of the English Navigation Acts.

Whilst the Scottish had been involved with tobacco from the days of its introduction to the British Isles, and had had the honour of supplying the world's first eloquent tabago-phobe, King James VI and I, they had been slow to build an international reputation as a tobacco-using nation. In general, they were well disposed towards the weed and King James was no more representative of his countrymen in his sexual habits than in his views on smoking. A Scotsman was quick to defend his country's honour on one count and William Barclay of Towie published *Nepthenes, or the Virtues of Tobacco* a few years after King James's famous *Counterblaste*. He was careful to exclude English excesses from his reasoned case for the weed, advising his readers: 'not, as the English abusers do, to make a smoke box of their skulls, more fit to be carried under his arm . . . than to carry the brain of him that cannot walk, cannot ride except the tobacco pipe be in his mouth'.

William Barclay also took pains to warn his readers off Spanish tobacco 'because either it is exhausted of spiritu-ality, or the radical humour is spent and wasted', and the Scots appeared to follow his recommendations of moder-ation and selection in consumption, at very least in their capital, Edinburgh. But amidst the heather and swirling mists of the Scottish Highlands tobacco took a stronger hold on the national psyche. Tobacco in general and snuff in partic-ular became an integral part of Highland custom. In a manner reminiscent of Amazonian Indians, a visitor to one of the tribes or 'clans' that populated this damp and isolated part of Europe would be welcomed with a pinch of snuff.

The adoption of the snuff horn as part of Highland costume has an origin as obscure as that of the kilt. It is likely that the snuff habit reached the Highlands from Ireland, with which the Scots maintained a regular trade, and where it had been observed as being well established as early as 1636: '[the] Irish take [tobacco] most in powder or smutchin, and it mightily refreshes the brain, and I believe there is as much taken this way in Ireland as there is in Pipes in England; one shall commonly see the serving maid upon the washing block, and the swain upon the ploughshare when they are tired with labour, take out their boxes and draw it into their nostrils with a quill and it will beget new spirits in them'.

The Scottish called the act of snuffing 'smeeshin', derived from the Gaelic *smuiden*, meaning dust. The Gaels were perhaps the first damp habitat race to prefer snuffing to smoking. It is likely that the choice was founded on practical considerations: it was near impossible to keep a pipe alight in the Scottish climate and the Highlanders instead focused their powers of invention on keeping snuff dry.

They achieved this aim partly through the evolution of a watertight container. In general, European snuff boxes were rectangular with a detachable lid. The Scots first changed the receptacle's shape to ensure the minimum surface area of its contents was exposed to the elements at any one time. This was achieved, like many of the best inventions, by adopting an everyday article, in this case, a ram's horn, to the purpose in hand. Their second innovation was to add hinges to the lid, which meant the snuff could be accessed without fully abandoning its protection against precipitation. Over time these 'mulls' became heirlooms and were passed on by clan chieftains.

The depth of tobacco's penetration into Highland culture can be measured by the legends that accumulated around it. One such, the *Scuttermull of Glen Moran*, first recorded in an eighteenth-century collection of Highland folk tales describes the appearance of the ghost of an ancient chieftain, Hamish Ich McFerr, at his successors' deathbeds. The apparition would be dressed for battle, his target on one arm, his wounds streaming blood. After materializing it was this spirit's custom to demand a smeeshin from the scuttermull. A pinch was duly dispensed and the scuttermull, along with various other hereditary clan items, was passed to the dying chieftain's successor. This Highland fable illustrates that, as with many other nations, Scottish tobacco use was attributed an ancience it did not possess.

Meanwhile, in the cities of the Lowlands, tobacco recommended itself to another aspect of the Scottish character. The trade's profitability appealed to Scottish merchants, who had helped the Virginian colonists to create new overseas markets for their tobacco, in particular in France where Orinoco had become so popular that the French neglected to plant the weed in their own colonies. Scottish traders developed into such a force in the tobacco trade that the Virginian collector of customs complained: 'I find that in these three years past there has not been above five ships trading legally in these rivers . . . above 20 Scots, Irish and New England vessels within these eight months have sailed out . . . with their loadings of tobacco . . . and the man of war has not discovered one of them.'

When the Navigation Acts started to throttle Scotland's trade, a process assisted by the frequent blockades imposed by the Royal Navy on foreign ports, its merchants lobbied strenuously for an exemption from their English neighbours, which was refused. Meanwhile, the Scots converted a trade crisis into a national disaster by attempting their first New World colony. Scots were already leaving for the Americas in numbers and success stories had filtered back of fellow countrymen who had made their fortune overseas, for example, an individual 'who was sent away by Cromwell to New England, a slave from Dunbar. Living now in Woodbridge, like a Scots laird, wishes his countrymen and his Native Soil very well, tho' he never intends to see it'.

Scotland's colony, named Darien, was founded on the
fever-ridden isthmus of Panama. The reasoning behind this
choice of location was that its strategic position, as a gateway
to both the Atlantic and Pacific Oceans, would enable
Scotland to dominate world trade. The Scottish people were
attracted by the logic of the scheme and committed their
savings to it. An initial body of 1200 colonists set forth,
and in November 1698 founded 'New Edinburgh', which
had grown into a cluster of mud huts and a fort by the time
it was abandoned the following year. Only 300 of the settlers
survived to return to Scotland. Darien virtually bankrupted
Scotland – the sum wasted on the venture represented around
a quarter of the country's total wealth.

The Scots' fellow Celts, the Welsh, had better luck in New
World trade. At various times in the second half of the seventeenth century England was fighting both the Spaniards and
the Dutch on either side of the Atlantic, and was forced to
subcontract some traditional naval functions in the
Caribbean to pirates, amongst whom Welsh blood was
disproportionately represented. These men had an appetite
for Spanish gold commensurate with the conquistadores' lust
for the substance in the prior century. Their hunger made
them useful allies for the English crown, which elected to
add legitimacy to their depredations. Henry Morgan, the
most famous Welshman in the New World until his premature death from alcoholism in 1688, was a paragon amongst
pirates. He was knighted by Charles II and appointed
Deputy-Governor of Jamaica in recognition of his efforts
against Spain. Morgan and his fellow pirates were furious
smokers, so much so that their tendency to smoke at all
hours and in all places was usually circumscribed by the
Articles of Piracy signed by a pirate upon joining the trade.
The Articles provided for distribution of plunder, compensation for loss of limbs in the course of duty, and as the
following example demonstrates, punished careless smokers:

ARTICLE 6: That any Man that shall snap his Arms,
or smoak Tobacco in the Hold, without cap to his Pipe,

or carry a candle lighted without lanthorn, shall receive
Moses's Law (that is 40 Stripes lacking one) on the
bare Back.

The pirates' part in England's official policy for the New
World suffered a terminal blow when their headquarters at
Port Royal in Jamaica, the largest English town in the New
World, was struck by an earthquake and sunk into the sea
in 1692, thus ending the prominent involvement of the Welsh
in seventeenth-century intercontinental trade.

By the 1690s the English under their Dutch king William,
Prince of Orange, had re-established supremacy in the colonial trade. Naval expenditure had leapt from around
£400,000 in the prior decade to over £2,800,000 in 1697.
Pirates were pensioned off or executed, and interlopers such
as the Scots were squeezed out of business. Impoverished
and oppressed by their powerful neighbour, Scotland's
merchants and its politicians pushed their country towards
union with England. The reasons for union, which resulted
in the creation of Great Britain, were principally financial.
Robert Burns, the famous smoker, lyric poet and pornogra-
pher, author of such gems as 'Nine Inch Will Please a Lady',
described his country as being 'bought and sold for English
gold' – a fair assessment of the combination of the two
states. As soon as union appeared inevitable, Scottish
merchants stockpiled tobacco at the lower Scottish import
rate, knowing that when taxes were harmonized between
the kingdoms they would be raised in Scotland. Many made
fortunes from this one way bet.

After England and Scotland became one country, Great
Britain, in 1707, Scotland's tobacco merchants were quick
to take advantage of their new British nationality. British
ships received the protection of the Royal Navy, now the
most powerful in the world. They were also able to exploit
their geographical advantage over merchants in what had
formerly been England: 'The Glasgow vessels are no sooner
out of the Firth of Clyde but they stretch away to the North
West, are out of the road of the privateers immediately, and

are often at the capes of Virginia before the London ships get clear of the Channel.'

The skills the Scots had acquired when outlaws greatly assisted them as legal traders. Instead of offering the colonists mining tools, olive presses or silkworm propagators in return for tobacco they brought slaves and gave credit. Within fifteen years of union they were responsible for 50 per cent of Great Britain's tobacco imports and controlled a significant proportion of the slave trade. The first fruit of their dominance was one of the larger, if uglier, tributes to Great Britain's tobacco habit: Glasgow, affectionately known by its inhabitants as 'Auld Reekie', developed into a major town on the back of the tobacco trade. The huts of salmon fishermen along the banks of the River Spey were displaced by tobacco wharves, from which ships would race away westwards towards the New World.

England's New World colonies had enjoyed relative peace while the various European powers battled for control of trade, and had prospered in the latter half of the seventeenth century. Settlers no longer needed to eat one another to stay alive; indeed, a sure sign of plenty, they had the highest birth rate in the Western world. Average population growth between 1660 and 1700 was 3 per cent against a European rate of less than 1 per cent over the same period. Life for Virginians was especially sweet. Foreigners bought their weed and slaves did the work. From the tentative beginnings made by John Rolfe in 1618 with an Orinoco crop of 20,000 pounds, Virginian tobacco exports had grown to 1.5 million pounds by 1640 and approximately 38 million pounds in 1700. Over the same timespan its population grew from 18,000 to 78,000 and life expectancy had long exceeded double figures.

The self-sufficiency fanatics to the north had also prospered. By 1700 they numbered 105,000 in New England alone and Great Britain's New World citizens constituted 10 per cent of its population, excluding slaves, who did not count as humans. Economic health in the north had derived from the sheer volume of shipping involved in the tobacco trade, which had encouraged other commerce to develop. An intricate web emerged, with tobacco at its centre. Tobacco was linked inextricably to almost every transaction between the colonies and the outside world. The northern settlers began to act as traders in their own right, specializing in commerce between the colonies. They supplied the Caribbean islands with tobacco, salt cod and agricultural products, receiving sugar in return, which they converted to rum which, in its turn, was used to purchase slaves in Africa. In the 1660s there were more than sixty distilleries in Massachusetts, producing 2.5 million gallons of spirits annually. The distaste for smoking the northern colonies had affected soon dissipated, and the often curious legislation, such as Connecticut's law which forbade anyone to smoke without a doctor's certificate and then not 'publickly in the street or any open places unless on a journey of at least ten miles', was repealed or ignored.

However, northern tobacco cultivation diminished. The farms on Manhattan Island in the Bowery and Greenwich Village areas became wharves, warehouses and conurbations. The intricate plots of jade green plants whose heady scent had refreshed New York on summer evenings were submerged beneath an agglomeration of planks, brickwork and sewers. The northern colonies viewed tobacco primarily as an item of trade – a means of gaining wealth, rather than wealth itself. This contrasts sharply with the position in the south, where tobacco so dominated the colonists' lives that it acted as their currency. It had served this purpose since 1616, but its use spread from 'big ticket' items such as wives and slaves to encompass everyday purchases. Tobacco could even be used to pay fines and taxes and a man's wealth was estimated in pounds of tobacco.

Unfortunately, the leap in tobacco production led to oversupply, a collapse in the price of the weed in the 1680s and the threat of economic ruin. As Lord Culpeper, Governor of Virginia lamented:

that which is more to us than other things put together,
and will be the speedy and certain ruin of the colony,
is the low price of tobacco. The thing is so fatal and
desperate that there is no remedy; the market is over-
stocked and every crop overstocks it more. Our thriving
is our undoing, and our purchase of negroes, by
increasing the supply of tobacco, has greatly
contributed thereto.

Maryland suffered the same damage as Virginia: 'Tobacco,
our money, is worth nothing . . . and [there is] not a shirt
to be had for tobacco this year in all our country.'

The collapse in the tobacco price affected not only tobacco
growers, but every colonist who kept their wealth in the
weed. The settlers did their best to maintain demand for
their 'money' at home and were observed by visiting
foreigners to smoke nearly as much as the English:

large quantities of it [tobacco] are used in this country,
besides what they sell. Everyone smokes when working
or idling. I sometimes went to hear the sermon; their
churches are in the woods, and when everyone has
arrived the minister and all the others smoke before
going in. The preaching over, they do the same thing
before parting . . . It was here that I saw everybody
smoking, men, women, girls and boys from the age of
seven years.

The catastrophe resulted in legislation in several colonies
aimed at maintaining and standardizing the quality of the
tobacco they produced. This succeeded in raising prices and
in restoring confidence to tobacco's monetary users. The
colonists had been quality conscious from the start. Indeed,
the first law in English passed by vote on American soil
concerned tobacco production standards and prices. But now
the laws were enforced in earnest and reinforced with further
legislation. By 1696, Virginia was able to pay salaries in
tobacco, including those of its clergy, who received 16,000

pounds of weed per annum. For the sake of comparison, if the same amount of tobacco could be sold on the UK retail market today, using the price of Golden Virginia hand rolling tobacco, the nearest modern equivalent, each priest earned £1,002,400 a year. These men of God were retained on such a stupendous stipend to 'supply this dominion with able and faithful Ministers whereby the glory of God may be advanced, the church propagated and the people edified'.

Once they had stabilized the tobacco market, legislators in Virginia, Maryland and Carolina then went a step further and established tobacco as a *de jure* as well as *de facto* currency. Tobacco notes were the first exchangeable instruments in the colonies and hence were precursors of the US dollar. Virginia led the way with its Tobacco Inspection Act of 1730. The Inspection Acts revolutionized tobacco regulation and the tobacco notes they introduced became a feature of domestic colonial trade for nearly a century. The Inspection Acts established public warehouses with official inspectors and required planters to transport every hogshead of tobacco in the colony to a warehouse for inspection. The inspectors were empowered to break open each hogshead, remove and burn any substandard weed, and issue tobacco notes to the owner specifying the weight and kind of tobacco. After 1730, Marylanders became aware that the inspection system gave Virginia an advantage over their colony by raising the quality and reputation of its tobacco. In 1747, the Maryland assembly passed the Maryland Inspection Acts which introduced similar measures to those prevailing in Virginia.

The tobacco inspection system worked as follows: if a planter turned in his weed 'loose' or in bundles he received a receipt known as a transfer note, which entitled the holder to a certain number of pounds of tobacco drawn at random from the total stock of transfer tobacco. Transfer tobacco was derived from several sources. It often happened that after filling his hogsheads, a planter had an insufficient quantity left over to fill another. This excess was delivered to the warehouse, where the planter would receive a transfer note in exchange. The clergy, and other colonists such as black-smiths and saddle-makers whose main occupation was something other than tobacco planting, often tended a small patch in their spare time in order to pay their taxes, and to make purchases in shops. These people carried their crops to the tobacco warehouse and received transfer notes that could either be sold or tendered as payment of debts, fees and taxes.

The reliance on tobacco to the extent that it could be trusted to do the work of gold demonstrated its pre-eminence in the southern colonies. A century after the weed had saved the citizens of Jamestown from starvation, it provided them with work or workers, refreshed their leisure hours, paid their priests and served as their currency. The richest man in Virginia, indeed, in all the British colonies was a tobacco farmer, Robert 'King' Carter, who owned over 300,000 acres of land and 390 slaves of working age.

Tobacco's importance was not limited to the colonies of the south. It was North America's principal export by value throughout the eighteenth century, representing nearly half of total exports in 1750. The weed had given the colonists a place in the world. The manner of its production, however, committed them to a social model which set them on a course of conflict with the metropolitan power, and ultimately, amongst themselves.

6

Magic Dust

Jobs for life: slavery in the British colonies
– tobacco production in the New World
and in Europe – tobacco in the Age of
Enlightenment – French lessons: the
snuffing craze – snuff spreads to Britain
and displaces smoking

Slaves had been a part of tobacco production since 1619 and their cost effectiveness, in working all their lives for free, recommended them above the indentured servants who had once constituted the backbone of its cultivation. They were imported to the American colonies in increasing numbers as the eighteenth century progressed. In the 1690s, Virginia's slave population was less than 10 per cent of its total; by 1750 it was 43 per cent and together with its neighbour Maryland it was home to over 144,000 slaves. The increase was sufficiently remarkable for a Virginian tobacco grower, William Byrd II, to complain to the colonial powers in Britain that 'they import so many Negroes hither, that I fear this colony will some time or other be confirmed by the name of New Guinea'. Byrd II further feared that the consequences might be a slave revolt that would 'tinge our rivers as wide as they are with blood'.

Hitherto, most slaves imported to British colonies had gone to the Caribbean islands where the received wisdom was that it was better to work slaves to death than

Eighteenth-century tobacco labels showing
slaves working in the tobacco trade

humouring them by allowing them to live to breed. For example, the slave population of Barbados increased by less than 25,000 in the first half of the eighteenth century, despite imports of more than 150,000. Slaves in the tobacco growing colonies were not considered intrinsically expendable to the same degree, and had the equivocal privilege of working with something with which they were familiar. The weed had been grown in west Africa prior to its introduction to Virginia, and although its cultivation was arduous, it was pleasant in comparison to the work of Caribbean slaves producing sugar, the new wonder crop.

Slaves on American tobacco plantations lived in compounds containing, on average, between five and twenty-five slaves. They worked in gangs to a tasking system, under which they were set objectives instead of working hours. Once the tasks for a day had been achieved their time was their own and could be used to clean their quarters, or complete minor tasks around the farm. The tasking system, and the fact that slaves were held in small units, enabled them a minuscule, yet invaluable opportunity to maintain and to perpetuate their culture, that in some cases was assisted by their owners' preference for slaves from a specific part of Africa. 'Callabars are not at all liked with us when they are above the age of eighteen. Gambias or Gold Coast are prefer'd to others. Windward Coast next to them', wrote a tobacco planter to his slave merchant when placing an order. Such discrimination created a degree of homogeneity on some plantations, enabling, at least, linguistic identity.

Despite the obvious limitations imposed on every aspect of their lives, slaves in the mainland colonies, in contrast to those of the Caribbean, were encouraged to reproduce, and their population became self-sustaining. Not only did American slaves breed, but despite the ever present threat of sale and separation hanging over families, they also succeeded in handing down a part of their culture to their children. Amazingly, family life was possible for slaves on tobacco farms. Slaves were seldom allowed to learn to read or write lest they spoil the fiction that they were not able to, so took particular care of their oral history. In many cases, the only keys to identity and culture that could be passed on were names. Early generations of slaves gave their children African names such as 'Quash' for boys born on a Sunday and 'Cuffee' for those born on Friday. While the names themselves survived the dislocation suffered by slave families their meanings often did not: they were simply symbols of continuity and resistance.

Slaves also developed new names unique to the colonies. Some of these were corruptions of African originals, while others represented the spirit of invention alive amidst repression. Whatever names slaves chose for their children, one was always automatically excluded from selection – that of their master. The need to control a portion of their own destiny also led to a general rejection of the pet names the slaves' masters presumed to bestow on their property such as 'Hercules' or 'Juno'.

The moral dilemma presented by slavery had not entirely escaped the colonists' attention. They were faced by the theological conundrum of whether slaves who converted to

Christianity might be kept in servitude. Jesus had exhorted his followers to love their neighbour. Could Christian love be compatible with bondage? The problem was solved, or rather bypassed, by legislation: 'all servants imported and brought into this country, by sea or land, who were not Christians in their native country . . . shall be . . . slaves, and as such be here bought and sold notwithstanding a conversion to Christianity afterwards'.

The involuntary additions to its workforce enabled Virginia and its southern neighbours to increase their tobacco production significantly during the course of the eighteenth century. Slaves were often a tobacco grower's greatest investment, as land was cheap, if not free, and the tendency of tobacco to exhaust the soil after two or three crops meant that planters moved frequently, taking their capital, i.e. their slaves, to the new areas selected for cultivation.

The other major tobacco producing countries in the Americas also relied on slave labour. Portuguese Brazil, the world's second largest tobacco exporter after Virginia, was its largest importer of slaves. In Brazil, conditions were so wretched that King James I, had he lived and travelled to see them, would have considered his *Counterblaste* justified, and the link between tobacco and hell proven.

Tobacco was also the fruit of many Spanish settlements, though none relied upon it to the same extent as Virginia. The Spanish crown attempted as tight a control over tobacco as its colonial rivals and its production was strictly regulated in all of Spain's American possessions. Some colonies were encouraged to produce the weed, whilst others were prevented. All tobacco production was subject to the royal monopoly, and all legal trade had to be conducted through the Tabacalera. As Spain strove to protect its colonial commerce a somewhat chaotic policy resulted and, for a time, tobacco cultivation was banned in Venezuela, its principal tobacco producer. Since tobacco was responsible for nearly half of Venezuelan exports, the country was effectively crippled. Fortunately, its captain general, Mexia de

Godoy, realized the futility of the prohibition and planters were allowed to cultivate on condition that they gave a guarantee that their crops would not be used for illegal trading.

Similar outbreaks of state paranoia affected Cuba, the second largest tobacco producer in the Spanish empire. When especially oppressive legislation – the factory and monopoly regime – was introduced by Spanish royal decree in 1717, the growers in Havana took up arms and rebelled on three successive occasions. This decree was one of history's most onerous and least successful attempts at tobacco regulation. Curiously, it relied upon King Leopold I of Austria's premise that tobacco was not an essential, and therefore a luxury that must be taxed. It is interesting to note that the same superfluity argument has never been advanced as justification for taxing say, tomatoes, nor any other single vegetable not strictly necessary to the survival of humanity.

Some of the tobacco produced in Spain's New World possessions remained in the Americas, where Spanish emigrants had finally succumbed en masse to the weed. Their clergy's demand for tobacco had become insatiable – even the shame brought upon the men of God by one of their brethren who, after taking snuff, had vomited a consecrated host on to the altar in the middle of Mass, could not dissuade priests from snuffing. The lay population meanwhile had taken up smoking, the habit so repugnant to their conquistador ancestors, although their weakness might be excused in part by their isolation. For much of the seventeenth and eighteenth centuries, Spain's American possessions had not been able to rely on official communication with the mother country, or, indeed, with each other. Beyond the reach of Spain's guiding hand, and with only Indians or slaves for stimulation, Spanish émigrés quietly adopted the smoking habit their predecessors had scorned as yet another instance of the Native American's innate inferiority. Tobacco's success invites the application of a biblical epithet to the conversion of expatriate Spaniards to its delights: 'the stone that the builders have rejected has become the corner stone'. Tobacco was assisted in its progress by its profitability. Whereas the Spanish Inquisition had banned the use of other Native American drugs, including peyote and coca because of their supposed satanic associations, they refrained from criticizing tobacco, knowing that the revenues it generated paid for their cloaks, hats and dungeons.

Isolation and a late conversion resulted in Spanish subjects overseas taking their pleasures in a different manner to other white men. Pipes were unheard of in Spain's South American empire. Gentlemen, and ladies, smoked *ceegars* (i.e. cigars), which in their turn were unheard of in England until 1735 when the traveller John Cockburn published an account of their use. Cockburn received his first *ceegar* from a group of Franciscan monks en route to Costa Rica: 'These gentlemen gave us some ceegars to smoke, which they supposed would be very acceptable. These are the leaves of tobacco rolled up in such a manner that they serve both for a pipe and tobacco itself. These the ladies, as well as gentlemen, are very fond of smoking; but indeed, they know no other way here, for there is no such thing as a tobacco pipe throughout New Spain, but poor awkward tools used by the Negroes and Indians.'

Most of Spanish America's legitimate tobacco production was destined for Spain. Once it had been packed by slaves it was shipped to Cadiz or Seville. Tobacco reaching Cadiz was rolled into cigars or ground into snuff, whereas the weed that arrived in Seville was processed at the new Fabrica del Tobacos, where it was made exclusively into snuff. The Fabrica's workers had been asked to produce 'Cadiz style' cigars in 1731, but had refused, claiming it was women's work. Besides, their hands were full – the Fabrica was working far in excess of its capacity.

The new Fabrica had been opened as a replacement for the St Pedro factory in 1687 during the reign of Carlos II, its construction part financed by his treasury. It soon outgrew its site, absorbing some of a neighbouring monastery and a series of charity dwellings dedicated to housing the poor. Demand for the snuff the Fabrica produced was soaring – in

1702 it had manufactured 1.1 million Spanish pounds, in 1722 nearly 3.1 million. In 1728 work began on a replacement for the second Fabrica that was to become the largest industrial building in the world. The third and final incarnation of Seville's tobacco factory was laid out as a walled city with its own chapel (presided over by its own Virgin patroness), its own laws and its own prison. It took nearly thirty years to build.

The financial logic behind this giant Fabrica was sound – European appetite for snuff had grown into a craze and the obsession needed feeding. A part of the surge in demand had a practical explanation. Smoking was still frowned upon in parts of Europe, and some German principalities had launched fresh campaigns against this form of tobacco use, focusing, in addition to godlessness, on the fire risks that smoking created. Snuffing was a less hazardous alternative. In Prussia, for instance, which had been a dedicated nation of smokers at the turn of the eighteenth century, a change in monarch resulted in a change of tobacco habit. King Friedrich I, followed by his son Friedrich Wilhelm I, had both been smokers and had held regular tobacco parties at which pipes were compulsory. Servants were dismissed, guests ate bread and cheese and drank beer, and the king and his best friend, ex-King Stanislas of Poland would sit and smoke together all night, sometimes consuming as many as thirty pipes each, a level of consumption high enough to send the most dedicated of tobacco shamans on a vision quest. In a further parallel to Amerindian tribal practice, matters of state were discussed informally, so much so that these parties became known as the 'Tobacco parliament' to which foreign statesmen would seek invitations in order to gauge the true sentiments of Prussia's ruling classes.

However, upon the ascension of Friedrich the Great, who did not smoke, the habit passed out of court circles. Friedrich disliked the smell of tobacco, considering its users fools, and in response to their ruler's prejudices, smoking was prohibited out of doors in several Prussian cities on the grounds that (a) it was slovenly, and demonstrated a lack of self-discipline, the possession of which attribute was there considered to be a virtue; and (b) because smoking was a fire risk. Both objections had appeared in other places at other times, but only in Prussia was the first made into a crime. The second, more reasonable excuse for oppression – that smokers caused fires – was pursued by Prussian officials with relentless fanaticism. Although Prussian streets of the age were certainly less flammable that the houses that lined them, smokers were confined indoors, to meditate their defiance over solitary pipes. Smoking was even prohibited out of doors in the countryside. Under another of Friedrich the Great's edicts against careless smoking, lighting up in harvest time was punishable by a month's imprisonment on bread and water.

If a Prussian wished to have a public life and a tobacco habit, he was reduced to using snuff. By the middle of the eighteenth century snuffing had taken hold of Prussia, to the consternation of social commentators:

> The world has taken up a ridiculous fashion – the excessive use of snuff. All nations are snuffing. All classes snuff, from the highest to the lowest. I have sometimes wondered to see how lords and lackeys, High Society and the mob, woodchoppers and handymen, broom squires and beadles, take out their snuff boxes with an air and dip into them. Both sexes snuff for the fashion has spread to the women; the ladies began it, and are now imitated by the washerwomen. People snuff so often that their noses are more like a dust heap than a nose; so irrationally that they think the dust an ornament, although, since the world began, all rational men have thought a dirty face unhealthy; so recklessly that they lose the sense of smell and their bodily health.

A second explanation for the eighteenth-century European snuff craze is to be found in the lofty realms of metaphysics. Just as Christian priests had adopted a habit associated with the enemy of their faith, so snuffing had spread from the nostrils of clergymen to those inhaling a more rarefied air in the salons and universities of Europe, whose owners set a higher value on life in this world than the men of God, and whose scientific and philosophical works were so revolutionary that the years in which they were produced were christened the Age of Enlightenment. This generation of thinkers, smokers and snuff takers was perhaps the first in Europe's history to throw off the bonds of superstition and to reject the concept that mankind was in decline, a philosophy that had persisted in various guises since the days of the Greeks, who had divided man's time on Earth into three ages: (1) the Age of Gold, when humanity had possessed giant stature, god-like perception, a life span of centuries and the ability to speak to animals, succeeded by (2) the lesser Silver Age in which some of these attributes were lost, followed by (3) a further decline, marking the beginning of the Age of Bronze where mankind found itself at present, dull and debased.

In contrast to this concept of decline, which the Catholic Church did much to encourage, hinting at centennial or millennial apocalypses, no less dreadful for being distant, a number of Enlightenment thinkers had suggested humanity actually might be improving, or, more than this, perfecting itself. In order to reach this conclusion, the prime movers of the Age of Enlightenment had investigated the heavens through philosophy and the Earth with their eyes, relying on deduction and observation as opposed to the inspiration of faith. They started by assuming one of the privileges hitherto accorded to God, and renamed the beasts of the field, the birds of the air and the plants of the earth. In 1735, Carolus Linnaeus, a Swedish tabagophile and Enlightenment figurehead, invented the science of taxonomy, a discipline by which every living thing on the planet could be ordered. Linnaeus gave the two species of tobacco plant new names, christening them *Nicotiana rustica* and *Nicotiana tabacum* in honour of Jean Nicot, Catherine de' Medici's court physician.

A willingness to meddle with the order of creation, and with superstition, was apparent in other fields beside those concerned with the animal and vegetable kingdoms. The

committed to selling pipe tobacco to the proletariat. In retrospect, at least in terms of freedom of expression, their work constituted tobacco advertising's silver age. Unhindered by consumer protection acts or health considerations, the scribblers employed hyperbole to promote their patrons' brands:

> Life is a Smoke! – if this be true,
> Tobacco will thy Life renew;
> Then fear not Death, nor killing care
> Whilst we have best Virginia here.

This slogan adorned the tobacco label of Leach Baldwin, a Greenwich firm, and was tastefully illustrated with a slave holding a tobacco roll. Interestingly, Grub Street's spiritual heirs – the modern advertising agencies – still base their operations in the same area of London and still pen jingles for the tobacco industry.

Part of the growth in snuff's popularity can be attributed to another fashion imported at about the same time to the British Isles from the continent: washing. The English had hitherto considered personal hygiene to be dangerous and the application of water or soap to the human skin was generally frowned upon. Despite the sneezes and occasional nasal discharges it provoked, snuffing was believed to be more sanitary than smoking. Smoking became associated with the olfactory dark ages from which society had just emerged and was condemned as a vice of the lower ranks, the clergy and academics. By 1773, Dr Johnson was able to observe: 'Smoking has gone out. To be sure, it is a shocking thing, blowing smoke out of our mouths into other people's mouths, eyes and noses, and having the same done to us; yet I cannot account why a thing which requires so little exertion, and yet preserves the mind from total vacuity, should have gone out.' The bulbous doctor, inventor of the dictionary, took snuff by the fistful and filled the pockets of his coats with the magic tobacco dust.

The social distinctions separating smoking and snuff

hardened to the extent that impenitent smokers were subjected to segregation and banished with their spittoons to areas where they might indulge their anti-social vice. The snuff revolution has its contemporary counterpart in the smoking and drinking habits of modern Britons, whose quest for purity has led them to reject the dark beers and heavy tobaccos preferred by their parents in favour of lagers and low tar cigarettes. It is fascinating to observe in habits of consumption the perennial supremacy of appearances, which altogether govern human perception.

As snuff displaced smoking in genteel circles, the English attributed to it many of the properties enjoyed by the prior form of tobacco consumption. A very large proportion of English snuff sales were medicinal. Superstition, it seemed, developed in pace with science and snuff was sold as the perfect stimulant for the Age of Enlightenment's newly postulated seat of reason. A typical example, Imperial Snuff, was claimed to be a cure 'for all disorders of the head, and . . . for all disorders of body and mind. It hath set a great many to rights that never was expected.'

The arrival of snuff was celebrated by the British government in its customary manner with red tape and taxes. Its attack commenced with an anti-adulteration bill, introduced by George I whose stated aim was to protect the nation's health and true intent to guard the sovereign's revenues. The bill limited the substances that could be added to snuff, specifically prohibiting wood, dirt and sand. Its real concern was that diluted tobacco meant less tax receipts, a portion of which went directly to the king. In addition to guarding its revenue, the British government also sought to extend its control over the domestic tobacco supply. In 1733, an excise bill was introduced to Parliament with the aim of creating a compulsory warehousing system for all tobacco imported into the UK. The excise bill provoked riots in the streets of London. Like so much of tobacco control legislation, it attempted to justify the extension of state control on the grounds that its citizens did not need tobacco.

The British, however, continued to resist the suggestion

The General P—S or Peace

that tobacco was a luxury. While they were prepared to differentiate between types of consumption on grounds of wealth, most of the United Kingdom's tobacco addicts believed tobacco, as their ancestors had, to be a necessity of existence. 'How long has tobacco been reckoned an article of pure luxury? Is it not good part of the subsistence of our poor? How many thousands in this city never taste a morsel of victuals till noon, nay very frequently till night, but a small dram and a chew or pipe of tobacco? To raise the price thereof . . . would therefore be as effectually starving these wretches, as laying a tax on flesh meat and bread corn.' In this instance tobacco consumers achieved a rare victory against legislators and the excise bill was abandoned.

The distinction between necessities and luxuries was one of considerable importance at the time, and subject to fierce political debate. Prevailing opinion held that poor people would be spoiled if allowed access to luxuries, so it was

addition to stargazing. After Tahiti, it was to proceed south
in search of a great continent reputed to exist beneath the
equator. Classical geographers had postulated that in order
to prevent the world from being top-heavy, its southern
hemisphere must possess an equal land mass to its northern
half. These undiscovered southern lands were tentatively
credited as being home to the wonders which the Americas
had been suspected of harbouring but had failed to reveal,
including the Fountain of Youth and the Garden of Eden.
They had been inked into various atlases, sometimes labelled
as Paradiso Terrestrialis – the 'earthly paradise', although
the title for the unknown lands while they eluded explorers
was also represented as Terra Australis – 'the land of the
south wind' – or Australia in English.

HMS *Endeavour*, navigated by James Cook, reached Tahiti
after a voyage of nearly eight months. Landfall was eagerly
anticipated by the ship's crew, who had spent the last month
living on salt beef and hard tack. Tahiti was a wonder.
Sapphire seas broke into snow white foam upon its fringing
reef, coconut palms overhung beaches of the finest sand, and
behind these was the Arcadia that had hitherto only appeared
in the imaginations of love poets. The Tahitians themselves
were in keeping with this idyll: scantily dressed in scarlet and
gold, the girls with flowers in their long hair, they spent their
days singing, making love and surfing.

Joseph Banks, the *Endeavour*'s naturalist, who spent much
of his time on Tahiti 'botanizing', was overcome by the scene
he encountered when he first landed: 'we proceeded for four
or five miles, under groves of coconut and breadfruit trees,
loaded with a profusion of fruit & giving the most grateful
shade I have ever experienced. Under these were the habi-
tations of the people, most of them without walls; in short,
the scene that we saw was the truest picture of *an Arcadia
of which we were going to be kings*, that the imagination
can form' (author's italics).

The Tahitians appeared equally pleased with their visitors, whom they entertained with feasts and dancing. The
young women offered themselves freely to the *Endeavour*'s

crew, officers and men alike. But the visitors carried the usual complement of infections, including various diseases to which the Tahitians had no immunity. They also brought intriguing and compelling habits – drinking alcohol and smoking tobacco that had as profound an effect on the isolated societies as the cats that they put ashore had on Tahiti's indigenous species, which had evolved in the absence of such predators.

The Tahitians were entirely innocent of tobacco, indeed had very little in the way of drug habits. They were, however, imitative of their European visitors and eager to become acquainted with their poisons, as the following excerpt from Banks's journal demonstrates:

> Tomio came running to the tents, she seizd my hand and told me that Tubourai was dying and I must go instantly with her to his house. I went and Found him leaning his head against a post. He had vomited they said and he told me he should certainly dye in consequence of something our people had given him to eat, the remains of which were shewn me carefully wrapd up in a leaf. This upon examination I found to be a Chew of tobacco which he had begg'd of some of our people, and trying to imitate them in keeping it in his mouth as he saw them do had chewd it almost to powder swallowing his spittle. I was now master of his disease for which I prescribd cocoa nut milk which soon restor'd him to health.

The *Endeavour* did not carry enough tobacco to habituate the Tahitians to its use. The ship's trade goods consisted principally of beads and iron nails, and sailors were careful to hoard their personal supplies of the weed. However, a precedent had been set and the process of global dispersal commenced by Columbus's fleet nearly three centuries before was about to be completed. After leaving Tahiti, Cook took the *Endeavour* on a looping course through the South Pacific, charting the entire New Zealand coastline, and discovering the east coast of Australia before returning to Britain. At each place he landed he made an impact on the peoples or environment he encountered, sometimes innocently. He and his crew have the honour of introducing tobacco to the last places on earth to receive the weed.

Not every civilization Cook had touched upon was untainted by tobacco. Tobacco had a twin in Terra Australis, or rather a tribe of relatives down under, as sixteen species of *Nicotiana*, the tobacco plant's genus, are endemic to Australia, including *Nicotiana gossei*, known as *ingulba* or *mingulba* to the Aboriginals, *Nicotiana excelsior* (*pulandu, balandu, pulantu, pituri*) and *Nicotiana benthamiana* (*muntju, tangungnu, tjuntiwari*). It is possible that the Australian Aboriginals were the first members of humanity to discover the deceivable weed, and in any event, had been using it for centuries before Columbus chanced upon America.

Aboriginal tobacco use followed a pattern already familiar from the Americas. It had both ritual and economic significance. The Aboriginal applications for the weed were practical, their reasons for consumption mythical and their methods ceremonial. The Aboriginals were sophisticated addicts. They varied doses to achieve different effects, applying the herb with a chemist's discrimination. Tobacco was used to induce trances at desert ceremonies, or ward off hunger when the migrant tribes crossed Australia's empty heart. It was prized for its properties as a stimulant and appetite suppressant – a vital quality for the far-ranging Aboriginals. It was also used before battle, to animate the warriors before facing their enemies. It was a symbol of friendship, and, finally, was an item of trade, serving as a currency that could be exchanged for weapons and tools with other tribes.

Aboriginals chewed their tobacco. Its leaves were dried and powdered, and sometimes mixed with the ash obtained by burning acacia wood. The ingredients were blended to form a brownish-grey paste, then rolled into cigarette-shaped cylinders. These quids were chewed, or stored behind the ear, where

their juices were absorbed transdermally. Such was the importance of tobacco that the Aboriginals, otherwise hunter-
gatherers, cultivated the plant. They kept no animals and grew
no food, but nonetheless would remain in a single place long
enough to harvest the weed they tended. However, in these
sparse lands, whose resources were quickly exhausted, there
was no time for an extended curing process. Instead, the
Aboriginals cured tobacco over fires which, although a risky
endeavour, helped to fix its chemical properties.

Restricted production and expedited curing led to a
limited supply of tobacco, which was traded over a giant
area of territory, where it acted as a currency between tribes.
The Aboriginals, as Cook observed, had few wants or
belongings, and hence possession of the weed conferred
status on its owner. Tobacco was one of the few exchangeable commodities, apart from weapons, the Aboriginals
possessed. A man with tobacco would, among other privileges, be granted access to the women of different tribes.

When an official account of the *Endeavour*'s voyage was
published in London by Dr John Hawkesworth in 1773,
Cook's discoveries caused a sensation. While the Europeans
had changed since the days of the conquistadores, being
faintly aware of the damage they could wreak upon other
civilizations, they still had a lingering suspicion that paradise existed somewhere on Earth and were convinced that
in Tahiti it had at last been discovered. Hawkesworth's book,
together with the lectures and journals of the *Endeavour*'s
botanist Banks, held Britain spell-bound.

The discovery of an 'innocent' or 'natural' society had a
profound effect on the enlightened, but still Christian,
Britons. On the one hand, they desired to offer these savages
the benefits of Christian salvation, and on the other, to
preserve them from Christian vices. A similar debate raged
on the other side of the English Channel. Unbeknownst to
Cook, Tahiti had been visited by a French ship under the
command of Bougainville, whose crew had been similarly
charmed by the Tahitians, whom they had infected with
gonorrhoea. Bougainville's naturalist, Philibert Commerson,

who, like Banks, had also 'botanized' on Tahiti, commented
that the Tahitians: 'know no other God but love'.
Bougainville's account of his voyages was published in
France at the same time as Hawkesworth's in England, and
it provoked as much debate as to the ideal human lifestyle.
French philosophers, led by Rousseau, had evolved a concept
of the 'noble savage', who epitomized the glories of
humanity, and who, it was argued, should be left to flourish
uncorrupted. Instead of condemning the Tahitians' civilization out of hand, the French compared its merits relative
to their own. They held their Christian monarchy up before
a looking glass, and dared to be disappointed with its reflection. This approach was summed up by Diderot, a disciple
of Rousseau's, who foresaw the damage Europeans might
cause to humanity in its natural state: 'one day they [the
Christians] will come, with crucifix in one hand and the
dagger in the other to cut your throats or to force you to
accept their customs and opinions; one day, under their rule
you will be almost as unhappy as they are'.

Alongside religion and alcohol, tobacco was also identified as a potential risk to the carefree Tahitians. It was
unusual that tobacco was held to be a threat. Whilst the
dangers of giving alcohol to savages were well documented,
tobacco had hitherto been absorbed into every savage society
to which it had been introduced by Europeans without any
apparent ill effects.

It seems that the objections to teaching new races to smoke
were largely moral. For the first time Europeans were willing
to consider the relative merits of the social model they had
evolved against those of other races. For the first time they
considered a response to innocence other than superior fire-
power.

But enlightenment did not yet rule European minds. Most
of its population relied on Christianity for their morals,
which called the Tahitian concept of free love the sin of
adultery, and which condemned the Tahitian practice of
strangling excess children as murder. Even the French, to
whom the concept of the noble savage had been a useful

allegory of the potential of a secular society, became tired of the South Pacific and its inhabitants. 'I endeavoured to stimulate their curiosity,' wrote Julien Crozet, a French lieutenant, who was killed and eaten in New Zealand, 'to learn the emotions that could be awakened in their souls, but found nothing but vicious tendencies among these children of nature.' The late Lieutenant Crozet's reflections on the inhabitants of the terrestrial paradise in the distant south seas suited the prevailing mood in France, whose population was starving and on the brink of revolution. Accounts of coconuts and dog feasts were no longer whimsical, and whether the inhabitants of New Scythera (as Bougainville had named Tahiti) still danced with their customary lewdness or sat slumped under their scenic palms surrounded by broken pipes and empty bottles was an irrelevance.

In England too the people had tired of foreign wonders. However beautiful and libidinous the Tahitians and their ilk might be, they were still distant cannibals who strangled their children, ate their dogs and performed human sacrifice. In London Banks was mocked for his association with savages and Dr Johnson added to the ridicule by questioning the scientific legitimacy of Banks's mission. Why he had bothered to bring back insects from a voyage around the world when England already had so many of them? The South Sea idyll collapsed when its consequences to the morals of Christian maids were considered:

> One page of Hawkesworth in the cool retreat,
> Fires the bright maid with more than mortal heat:
> She sinks at once into the lover's arms
> Nor deems it vice to prostitute her charms;
> 'I'll do,' cries she, 'what Queens have done before';
> And sinks, from principle, a common whore.

Amidst this general atmosphere of disillusionment with all things primitive, one of Cook's discoveries – Botany Bay in New Holland (as the country of Australia was then known, in recognition of the discovery of its west coast by the Dutch

their religion, imported reading material and their own publications.

The paranoia generated by such conspiracy theories was overwhelming, and as a consequence, every act of the British government was viewed with the utmost suspicion, a suspicion that was justified by the level of idiocy present in the government of King George III. Snuff and fashion still ruled British society, brightly coloured uniforms motivated its armed forces. Tea had displaced brandy as the beverage of preference amongst the traditionally hard drinking British. Life in London was considered too fascinating and exciting to listen to soberly dressed colonials, whom, it was rumoured, had taken to chewing tobacco like animals. This indifference was nowhere more evident than at society's pinnacle, represented by the monarch. George III had so little interest in the world outside London that he did not even see the sea that encircled his kingdom until of advanced years and mad. His wife, Queen Charlotte, was known as 'snuffy Charlotte' as snuffing was what she did most and best.

The colonists, meanwhile, had started to formulate ideas of self-government, based on the premise that a country could exist without a king. Many of the colonists were familiar with the classical brand of republicanism, having encountered the theme when perusing Roman texts in search of civilized precedents for justifying slavery. They also revived another Roman notion: patriotism. The fruits of the colonists' enlightened speculations on classical principles were disseminated in pamphlet form by countless printing presses. The printers gave their work for free, in protest against a new British tax on printing. Some of these publications reached London, where they were read with exasperation. It seemed that not only were the colonists avoiding taxes, but proposing that they should not pay them ever.

The matter quickly became one of principle too for Great Britain. A firm hand was needed: 'The Americans must be subordinate. This is the mother country. They are children. They must obey, and we must prescribe.' In 1770, two regiments were sent from England to Boston, where they were to be quartered at the colonists' expense. When the settlers protested against this imposition by snowballing the British troops, whose trademark redcoats made tempting targets, the soldiers responded with bullets, killing four. The event was magnified into a massacre by the colonial printers. Thousands of copies of a reconstruction etched by Paul Revere were distributed among the disenchanted. Various trivial incidents were further catalysts to hostilities. In Boston, in 1773, the surplus tea Britain had shipped to its colonies in the hope of addicting them to this invigorating herb was cast into the harbour by white men dressed as Indians. The British were horrified. This was not merely insolence, but uncivilized insolence. Dr Johnson, who had an opinion to hand, commented: 'Patriotism is the last refuge of a scoundrel.'

But the high-minded colonial agrarians were not occupied merely with principles and the revival of Roman ideals. They were also motivated by interest, a term used by George Washington to represent the personal matters that committed men to action: 'Men will speculate as they will, they may talk of patriotism . . . but whoever builds upon it as a sufficient basis for conducting a long and bloody war will find themselves deceived in the end . . . for a time it may of itself push men to action, to bear much, to encounter difficulties, but it will not endure unassisted by interest.' In Washington's case, his 'interest' was that payable on his debts to British tobacco merchants.

Although the progress towards fighting was slow, this was no more than a reflection of the times. Colonial orders travelled under sail and took months before they could be acted on. The colonists did not have to wait for a ship to arrive before acting, and their cause gained momentum whilst the British were paralysed between moves. Their conflict with their parent was a media war, in the sense that the colonists' hearts and minds were won over to the cause not so much by British brutality but by the perception of it. Murders were transformed by the colonial printing presses into genocide, the theft of a chicken by a redcoat into a personal tragedy whose pathos resonated in the breast of every reader.

Thomas Jefferson, the eloquent Virginian tobacco farmer, did much to formulate the colonists' arguments in his treatise, *Summary View of the Rights of British America*. The treatise focused on a novel concept – human rights: 'The God who gave us life gave us liberty at the same time: the hand of force may destroy but cannot disjoin them.' The treatise was considered amusing reading in London. The colonists' persistence in living among their slaves gave a surreal edge to their philanthropic speculations. Dr Johnson again had a view: 'how is it that we hear the loudest yelps for liberty among the drivers of negroes?' The British government's response was to send more redcoats and suspend the Massachusetts assembly. They still believed they were correcting wayward children, and it was the colonists who elevated the disagreement into a war of sovereignty.

The thirteen British colonies drew up a Declaration of Independence, by which they formally assumed control of their own sovereignty. A majority of its signatories were involved in the tobacco trade. If one were to search for similarities among the fifty-six men who drew up the birth certificate of the most powerful nation in history, one would discover a belief in God, and tobacco interests. The Declaration began with a preamble enumerating the colonists' grievances against Britain, many of which were inspired by tobacco, the foundation stone of their overseas trade and hence national identity. The Declaration was adopted by the thirteen rebel colonies in July 1776, and announced to the world by the United States – the collective identity they had chosen for themselves.

The men with the best idea of the likely progress of Britain's war against the United States were Glasgow's tobacco barons, who had become the kingdom's principal tobacco traders, importing 49 million pounds out of a total of 90 million in 1772. The tobacco barons inhabited huge houses in what is now Glasgow's commercial centre and were given to wandering the docks dressed in scarlet cloaks with gold laced hats. They were familiar with true conditions in the New World, and just as they had done before union with England, began to stockpile tobacco as soon as they had news of the Declaration of Independence. By the outbreak of war, Glasgow's tobacco warehouses were full and thirty-two ships lay unloaded in the harbour. The tobacco barons' prudence was justified as the price of the weed rose from three pence per pound to thirty-eight pence. The shortage of the weed led desperate British subjects to reintroduce home-grown to Yorkshire and Scotland. The redcoats destroyed their crops and imprisoned the growers, and were more successful in their persecution of tobacco cultivators in Britain than the United States. The British had to rely for their supplies on captures in order to relieve the suffering on the home front. Between 1776 and 1782 the United States exported 87 million pounds, of which 34 million were captured by the Royal Navy.

Contrary to British expectations, the colonists could fight as well as formulate declarations. While the British usually prevailed in set-piece encounters, the United States adopted a guerrilla mode of warfare that, in turn, was used so effectively against them almost two centuries later in Vietnam. They also organized an independent and coherent administration, which raised taxes and established diplomatic relations with foreign powers, especially those hostile to Great Britain. This combination of the sword and the pen, of guerrilla tactics and state visits resulted in an alliance between the rebels and France in 1778, and with Spain in 1779, whereafter the tobacco war proceeded slowly but inevitably towards a victory for the colonists. While the French were enchanted by the appearance of Benjamin Franklin, who represented the United States abroad: 'His clothing was rustic, his bearing simple but dignified, his language direct, his hair unpowdered. It was as though the simplicity of the classical world, the figure of a thinker of the time of Plato, or a Republican of the age of Cato or Fabius, had suddenly been brought by magic into our effeminate and slavish age,' they were, however, wary of the Republican cause – as King Louis XVI observed: 'it is my profession to be a royalist'. Such qualified sympathy led the French to require financial consideration for their participation, which was supplied by tobacco. Five million pounds' weight of the weed acted as collateral for the loan Benjamin Franklin won from Europe's most potent nation. In this manner, tobacco supplied both the cause and the victory in the United States War of Independence.

Upon learning of news of this unnatural conjunction with the French, the British turned their energies towards destroying the security upon which the commercial aspects of the alliance were based. In addition to attacking the rebels' shabbily dressed armies the redcoats under Lord Cornwallis launched assaults against their life blood: the Virginian tobacco crop. The fields were fired, the curing barns burned down, and slaves were sometimes liberated. These last had proved less eager to fight against their masters than the British had anticipated, perhaps because some liberated slaves were considered best able to serve Britain's war efforts by joining other slaves on sugar plantations in the loyalist Caribbean islands. In between vandalizing tobacco farms, Lord Cornwallis manoeuvred his army into encirclement at Yorktown, where the combination of a French fleet and United States soldiers forced him to surrender. After giving up his sword he might have reflected that the substance that burned in the pipe offered to him in exchange for his weapon was the fuel that had consumed his monarch's dream of subjugation.

The surrender caused consternation in London. How could British trained mercenaries (most of Cornwallis's army were Hanoverians) surrender to a collection of tailors and tobacco farmers? Lord North's government resigned and the new administration sought to end the conflict. A meeting was arranged in Paris in 1782 where Britain and the United States met to negotiate a peace. Much to the amazement of the French delegates, the peace was concluded quickly and in a surprising spirit of amity. The British had recently annihilated the French European and Caribbean fleets, defended Gibraltar against the Spanish, triumphed in India, discovered Australia and felt that their pride had been restored in some measure. Tobacco debts were negotiated as part of the reparations and trade in the weed resumed.

A key to the goodwill between Britain and the United States can be found in the exchange between an aristocratic French delegate to the treaty processes and Caleb Whitefoord, a British representative. When the aristocrat predicted that the 'thirteen United States would form the greatest empire in the world', Whitefoord responded: 'Yes Monsieur, and they will all speak English, every one of them.'

Within ten years, many of the French delegates at the Treaty of Paris were speaking English too. The expenses of the last war, and the enormous running costs of the French court, led to intolerable pressure on the French peasantry, whose tax burden was double that of their counterparts in Britain. The most hated of these impositions were the five great 'Fermes' or general taxes on salt, spirits, customs, market stalls and tobacco. The French peasantry were regular smokers, relying on tobacco's appetite suppressant properties to aid them through regular famines. Encouraged by the success of their allies in the United States they revolted, and established a republic of their own. Snuffing almost vanished in Republican France. The habit was indelibly tainted by its association with courtiers and kings, and in the early, guillotine happy years of the French Revolution, even committed snuffers were forced to switch to smoking lest a dust-covered lip or a bad sneezing habit might cause them to lose their heads.

For the first two years of liberty, the French partisans were the only citizens of a European country to enjoy tax free tobacco. This could not last – the revolutionaries had inherited a bankrupt state and a tobacco tax was the easiest way to replenish its coffers. Commencing on 29 January 1791, a momentous debate was held at the National Assembly in which principle was ranged against necessity. Mirabeau, the great statesman of the Revolution, argued for a universal tobacco tax, though at a lower rate than that imposed by the hated Fermes. He was opposed by the deputy M. Roderer, on the principle that any tobacco tax would be a crime against the poor, and an insult to the Revolution. Roderer proposed that every partisan be empowered to grow, manufacture and sell tobacco tax free if they so desired. The result was a compromise – a light, patriotic tobacco duty which few French partisans paid.

Fortunately for the state of French public finances an individual came to power who was prepared to enforce tobacco taxes. Napoleon Bonaparte, commander of the French army in Italy, victor against Austria, had installed himself as one of three consuls entrusted with governing France in 1799. By 1807 he had attacked and defeated the Austrians, Prussians and Russians and had dismembered the Holy Roman Empire which had existed as a state since AD 962. Napoleon's personal tobacco habit was snuffing, the favourite pastime of the *ancien régime*. He took a kilo of snuff each week, the equivalent to a hundred-a-day cigarette habit. Napoleon collected snuff boxes. He had expensive tastes and his snuff boxes were usually made from precious metals or ivory, amber and tortoiseshell, and decorated with pictures or medals of his role models, who were other tyrants and autocrats such as Julius Caesar and Alexander the Great. His favourite, which he believed to be a lucky snuff box, bore a portrait of his first wife Joséphine de Beauharnais on its lid.

Napoleon enforced French tobacco taxes, which soon were generating more income than they had when France was ruled by kings. In addition to redrawing the political map of Europe, his armies were responsible for profound changes in the continent's tobacco habits. The fear and famines they occasioned were in part responsible for a surge in demand for the weed; the alterations they made to established social structures also enabled the practice of smoking to spread to nations which had hitherto condemned it, or neglected it in favour of the eighteenth-century craze for snuff.

8

Bandoleros, Scum and Dandies

Tobacco during the Napoleonic Wars –
Spanish bandits introduce the British to
cigars – the liberation of Spain, and Europe
– smoking returns to fashion in Britain –
Prussians scrutinize tobacco
and discover nicotine

After Napoleon had subdued continental Europe, only Britain stood against him in the Old World. Protected by its navy which had destroyed a combined French and Spanish fleet at the battle of Trafalgar in 1805, Britain was impossible to invade. Napoleon therefore applied a continental blockade which prohibited any of France's subject nations, or 'allies', from trading with Britain. Russia, which had aligned itself with Britain until its defeat at the battles of Austerlitz and Friedland and its subsequent forced friendship with France, elected to ignore the blockade, and Napoleon resolved to enforce it by the sword. He mobilized a giant army to invade Russia, which in turn prepared for war. Tolstoy captured the mood of the Russian nobility in his novel *War and Peace*, the following excerpt from which shows the aristocracy gathered together at the house of Count Rostov, where they smoke his 'connoisseur's collection of Turkish pipes' and discuss the coming conflict.

In the count's room, full of smoke, there was talk of
the war, which had been declared in a manifesto, and
of the levies of troops. The manifesto no one had yet
read, but everyone knew of its appearance. The count
was sitting on an ottoman with a man smoking and
talking on each side of him. The count himself was
neither smoking nor talking, but, with his head
cocked first on one side and then on the other, gazed
with evident satisfaction at the smokers, and listened
to the argument he had got up between his two neighbours.

Russian smoking habits were Asiatic at the time. They used
tobacco exclusively for leisure, and took it from narghiles
with amber mouthpieces. Smoking was predominantly a rich
man's diversion, although as Napoleon commenced his
attack and pressed into Russian territory, common soldiers
took up the habit. As had been the case in the Thirty Years
War nearly two centuries before, the strain of battle led to
a leap in demand for tobacco, whose calming properties
appeared designed for fighting soldiers. Narghiles were not
convenient for campaigning, and so other smoking devices
came into use among Russian troops, including cigars, whose
popularity followed in the wake of Napoleon's armies,
thanks to a supply from their allies in Spain.

Spain was a weak spot in the continental blockade. British
ships were able to touch on its coasts at will. Napoleon
perceived an opportunity to strengthen the blockade when
his Spanish allies were distracted by a succession battle over
their throne, and decided to introduce his brother as a candidate. His generals settled the contest in his sibling's favour,
the Spanish discovered they had a new king, and Joseph
Bonaparte settled down to rule a country occupied by
France.

One of the French armies necessary to ensure that the
Spaniards respected their new monarch was based in
Andalusia where they were able to enjoy the creations of
Seville's Fabrica del Tobacos. The Fabrica's range of products

had expanded from its eighteenth-century staple of snuff to include three types of cigar, and a form of raw tobacco, called 'tobacco picado'. Its new products were the consequence of a change in Spanish tobacco consumption patterns. Little by little, the colonial habit of smoking was superseding snuffing.

Although Spain's clergy were steadfast in their commitment to snuff as the most discreet form of tobacco addiction, and possibly because of the potentially detrimental effects on their dignity of a flaming cigar clamped between their jaws, the country's moneyed and leisured classes had been converted to the cigars enjoyed by their colonial counterparts, which they smoked openly – any stigma attached to the cigar's pagan origins had been forgotten. Their consumption was further encouraged by the ancient Andalusian sport of bull fighting, which had become a fashionable spectacle, and at which a cigar was obligatory for any hidalgo of blood. The three varieties of cigar the Fabrica produced to cater to this evolving market were graded by size and cost. The smallest and cheapest were Cadiz style, followed by Papantes, which were of a medium size, and whose ends were tied off with red thread, as opposed to the white used for Cadiz style, and Puros, the largest and most expensive of the Fabrica's range, preferred by the elite, including King Fernando VII, whose absence had created the vacancy now occupied by Joseph Bonaparte.

The officers in the French army occupying Andalusia were delighted by Puros, the most distinguished of the Fabrica's offerings, while their men became equally enamoured of Papantes and Cadiz style. They ignored altogether the final category of product made at the Fabrica – 'tobacco picado' – literally 'minced tobacco', which was left to its traditional market amongst Spain's poor. As tobacco use had become endemic in Spain, its largely rural population had selected smoking as their habit of preference, and in imitation of the Aztecs their ancestors had conquered consumed their tobacco wrapped in a twist of maize husk. City dwellers substituted paper for leaves and the device acquired a name: 'papelote'. An early example of a *papelote* appears in Goya's 'La cometa' (1778).

The *papelote* was a versatile way of smoking tobacco. When tobacco supplies were short they could be eked out by rolling slender *papelotes*; if the fine Valencian paper smokers preferred was not available a leaf from a book, or a piece of newspaper, might serve as substitute. Since each man was his own manufacturer there were as many different styles of rolling *papelotes* as there were pairs of hands. Passionate and sometimes violent arguments are recorded as to which was the best method of wrapping tobacco in paper.

The paper wrapper also made the smoker master of its contents. The *papelote* might hold shredded cigars mixed with snuff, it might be filled with tobacco picado, it could also hide entirely different, possibly illicit substances. Ever since the United States had won its independence, contraband tobacco had flowed into Spanish harbours, either directly, over the Atlantic, or latterly, via British ships running Napoleon's blockade. The colour of the pale Virginian leaf contrasted with the dark Cuban and South American tobaccos that made up official imports and gave it away at once as smuggled. Only when swaddled in paper could the illegal weed's identity be concealed. The Spanish government, sensible of lost revenues, had struck back at this threat to its monopoly. In 1801, it had enacted a law in the king's name which lamented the ingratitude of his subjects, for whose pleasure tobacco factories had been specially constructed, and forbade the consumption of 'white', i.e. paper wrapped, tobacco. The new prohibition was additional to existing legislation in support of the royal monopoly, under which growing, importing, selling or buying foreign tobacco was punishable with fines, beatings and imprisonment. The 1801 legislation was ineffective, and in 1802 further prohibitions were issued, which carried punishments of increased severity: any person caught selling 'white tobacco' was to be whipped, if a state employee was found smoking it he would lose his job, if a peon or countryman was spotted with a *papelote* in his mouth he faced the option (at the judge's whim) of banishment or two years' hard labour, and if a woman transgressed, she was entrusted to a convent, or a lunatic asylum, for up to four years.

The arrival of French rulers and French armies in Spain produced a boom in contraband. The shortages they created in Seville itself resulted in internal smuggling. Fabrica workers, nicknamed 'tarugos', carried the dusky government tobacco out of the factory a kilo at a time, concealed in their rectums. But these altruists alone could never satisfy the excess of demand over supply that prevailed in Spain. In the turmoil occasioned by the French occupation extensive smuggling chains developed to carry necessities such as cloth and tobacco from Spain's coast to its heart. However, the social functions performed by smugglers in nineteenth-century Spain were overshadowed by the romantic deeds of the practitioners themselves.

Bandoleros, as these men became known, deigned to use their rectums for tobacco smuggling. They waded chest deep into the surf to land hogsheads of Virginian weed, whose contents were transported throughout the Iberian peninsula. Their deeds were celebrated in song and verse and whenever one of their number distinguished himself by his valour or brutality he acquired a nickname. These varied between the poetic and the prosaic. For instance, Jose Maria was known as 'El Tempranillo', or the 'early one', on account of his tendency to land tobacco or surprise his victims in the mornings, whereas Luis Munoz Garcia revelled in the title of the 'Sponge-cake of Borge'. As the bandoleros organized themselves against the French, they adopted collective *nomes de guerre*, such as 'Los Siete Ninos de Ecija' the 'seven sons of Ecija', who were later to win glory by intercepting a consignment of cigars and treasure sent as a gift to King Fernando VII from Cuba. Bandoleros took pride in scorning base afflictions like hunger and cold. The only cravings they would admit were noble – wine, women, poetry and tobacco

– these were the things a true man needed to sustain his patriotism, or avarice.

Their theatres of operation were the immense plains, moors, forests and jagged mountains that separated Spain's cities and offered scenes of isolation and desolation unrivalled in the rest of Europe. These gentlemen of the sierras were responsible for the evolution of much of latter day smoking philosophy. Contemporary cigar smokers are indebted to the bandoleros for the immortality of the gesture and the inseparable association of tobacco with the concept of 'machismo'.

Moustaches were much in vogue among the bandoleros, alongside cartridge belts which they slung from their shoulders across their chests, full length capes, leather boots and wide brimmed hats. They rode like demons, and, when conditions were favourable, were mounted on semi-wild Andalusian stallions, which would only respond to their masters' commands.

The bandoleros were joined in their resistance to the French by a British expeditionary force which was dispatched to Spain in 1808 following a surprise alliance between the two countries. It had a disastrous start, losing its commander, Sir John Moore, who was killed and buried in his boots at the siege of Vigo. His replacement was Arthur Wellesley, whose battle experience in India counted against him in Whitehall where he had been contemptuously referred to as a 'Sepoy general'. What Wellesley lacked in social graces he made up for in tactical skills. His army, which he described as the 'Scum of the earth' after they had sacked Vigo and burned and pillaged Coruña, inflicted a series of defeats on Napoleon's generals in a campaign that lasted between 1808 and 1814. As in so many other conflicts, the pressures of battle caused soldiers, officers and men alike to resort to their patron saint: tobacco.

At the beginning of the campaign, smoking was more prevalent amongst French troops, whose proximity the British claimed to be able to detect by 'the smell of baccy and onions'. However, as the campaign progressed, the British soldiers and their officers became as ardent devotees of smoking as their enemies. Wellesley's troops suffered extreme privations of hunger, and the utility of tobacco as an appetite suppressant was in part the impulse behind its popularity. As the following excerpt from a private soldier's diary illustrates, smoking inured the British force to the hostile conditions in which they fought: 'Weather unfavourable and harassing. Winter fast approaching. On piquet every other night we make large fires and contrive to stupify ourselves with brandy and tobacco.'

British troops obtained much of their tobacco from their Spanish allies, usually in the form of cigars, which they found easier to use on the march or when in battle than their traditional clay pipes. Tobacco was also shipped from England along with other rations for the troops, then carried by mule train to the front. Wellesley's military genius was not restricted to combat – he was an outstanding organizer and his troops, though often hungry, did not endure starvation like their counterparts in the French army. When not fighting, relationships between opposing troops were friendly and the British often traded their tobacco supplies with the enemy for brandy: 'At this time, and for a great part of that in which we were quartered here, a very friendly intercourse was carried on between the French and ourselves,' wrote Edward Costello, a British officer at the lines of Torres Vedras.

Tobacco was in great request; we used to carry some of ours to them, while they in return would bring us a little brandy. Their 'reveille' was our summons as well as theirs, and although our old Captain seldom troubled us to fall in at the 'reveille', it was not unusual to find the rear of our army under arms, and, perhaps, expecting an attack. But the Captain knew his customers, for though playful as lambs, we were watchful as leopards.

The leopards triumphed. The French were driven from Spain and engaged and defeated on their own soil. The British victory in the peninsula coincided with Napoleon's defeat in Russia, where he lost 490,000 of his Grande Armée's 500,000 men. His response to catastrophe was offhand: 'All that has happened is nothing . . . it is the effect of the climate, and that is all. From the sublime to the ridiculous is but a small step . . . I am going to raise 300,000 men.' Before he had the opportunity to fulfil this vow, Napoleon was exiled to the island of Elba and the leaders of the Russian, Prussian and Austrian alliance that had defeated the 'Corsican Brigand' gathered in London to be fêted. As a consequence, smoking reappeared in Britain at all levels of society. While the poor had never abandoned the habit, the rich resumed it. In addition to the examples set by returning officers, society was influenced by visiting dignitaries, including the Prussian General Field Marshal Prince Blucher von Wahlstatt, an ardent pipe smoker, who was seldom seen without his favourite two foot tube that hung down vertically from his lips to his navel.

Prior to the Peninsular War, snuffing had held on to the position that it had acquired in the eighteenth century as the pre-eminent British tobacco habit. At the overcrowded gatherings of the affluent classes it served a useful social function as a substitute for speech. The manner in which a pinch of snuff was disposed of was a mime show of the snuffer's emotions, which it was then considered otherwise bad form to reveal. Snuffing's popularity had survived changes at court, where King George III's insanity had progressed to the extent that even by the standards of his time, he was considered unfit to appear in public. He had therefore been confined and his constitutional responsibilities devolved to the Prince of Wales, nicknamed 'Prinny', an overweight libertine whose full figure attracted unkind comments from his subjects, upon whom the coincidental resemblance between the Prince's title and marine mammals was not lost:

By his bulk and by his size
By his oily qualities
This (or else my eyesight fails)
This should be the Prince of Whales.

During the term of Prinny's regency, British society underwent a revolution, though of a very different nature to its
continental counterparts, in the course of which every aspect
of the old order, bar the snuffing habit, was overthrown. This
revolution's figurehead was George Bryan Brummell, aka
'Beau' Brummell, described as a man 'heroically consecrated
to this one object: the wearing of clothes wisely and well;
so that as others dress to live, he lives to dress'. Beau
Brummell was an English version of Casanova who wooed
men instead of women with his charms. Much of Beau's
fame derived from his fastidious dedication to personal
hygiene which, while it amazed his contemporaries, might
not seem excessive today. Beau washed every day, changed
his clothes every day, had regular hair-cuts and dieted and
exercised to maintain his figure. He caused a sensation at
court, which had never before witnessed anyone quite so
clean, and where people often lost their looks and forgot
their figures in their twenties.

Beau Brummell's revolution occurred in men's clothing,
and the wigs, ruffs, embroidery and britches that had been
more or less court fashion for the last 200 years were abandoned for what is recognisably the precursor of modern
dress. Brummell's innovations were aided, in part, by a hair
powder tax the British government had attempted to impose
on the nation, which led to a boycotting of wigs and protest
hair-cuts in the House of Commons. Short hair suited Beau's
new look, whose other motifs were dark, tailored coats,
spotless white cravats and long trousers held straight by
straps under the heels of black leather booties. Its philosophy was a kind of minimalism: simplicity of line, quality
of materials and perfectly executed tailoring. The new
austerity had its drawbacks: 'Dandies', as Brummell's
followers were called, often cut their ears on their starched

cravats if they attempted to turn their heads sideways.

The emphasis on cleanliness led Beau to revise snuffing to ensure less spillage, and to introduce emission control by phasing out sneezing. Both his taste in the powder and mannerisms while consuming it were widely imitated. Londoners had become as modish as the Parisians Casanova had observed fifty years before. Brummell's favourite blend – 'Martinique', a light, unscented concoction – was supplied by Friburg and Treyer, London's oldest and most exclusive snuff shop, founded in 1720. Beau converted Prinny, the Prince of Wales, to snuff, who also patronized Friburg and Treyer, and this association was announced on their labels as the first royal endorsement of a tobacco product.

Short hair and neat clothes were not the only influences at work in British society during the Regency period. The simple formality of men's clothes was reflected in Georgian architecture and equally simple, if revealing, women's fashions where the layers of petticoats of the prior century had been abandoned, as indeed had all underclothing. Women also continued to take snuff, following the example of Queen Charlotte, whose favourite blend was 'Morocco', which she bought, twelve pounds at a time, from the same supplier as her son and heir.

This apparent unity spawned a counterculture, known as the 'Romantic' movement, whose writers and painters worshipped the wild and untamed aspects of nature, with which they encouraged human interaction at the most primitive levels. Its initial prophets, Samuel Taylor Coleridge and William Wordsworth, both pipe smokers, enchanted society with their poetry, which it interpreted (to their exasperation) as charming Arcadian lyricism. The Romantic writers made society receptive to simple, pseudo-rustic activities such as building follies, spending time out of doors, and smoking.

A second wave of Romantic writers further advanced smoking in society's perception. At its crest was George Gordon, Lord Byron, a poet determined to put the sex back into romance and whose credentials, which included a title as well as genius, gave him automatic access to society's highest strata. When Lord Byron arrived in London, his wit, charm and extravagantly good looks caused a sensation, and his behaviour several scandals. Byron's presence in society appeared at first to stimulate the snuff habit. The Duke of Sussex, married to Augusta Leigh, the first of Byron's affairs, and coincidentally his half-sister, opened an account at Friburg and Treyer halfway through his messy divorce proceedings (which then, in the case of peers, required the individual sanction of the House of Lords). His snuff consumption rose as the case proceeded.

Byron was an unashamed exponent of smoking. He had picked up the habit abroad in the course of extensive travels and had observed the hold tobacco had over different faiths and nations. His work enjoyed the popularity its excellence merited, and when Byron celebrated his favourite method of consumption, with perhaps the most triumphant commemoration of tobacco ever written, cigars began to appear between the lips of his London acolytes:

> Sublime tobacco! which from East to west
> Cheers the Tar's labour or the Sultan's rest;
> Which on the Moslem's ottoman divides
> His hours and rivals opium and his brides;
> Magnificent in 'Stamboul, but less grand,
> Though not less loved, in Wapping or the Strand:
> Divine in Hookahs, glorious in a pipe,
> When tipped with amber, mellow, rich and ripe;
> Like other charmers, wooing the caress,
> More dazzlingly when daring in full dress;
> Yet thy true lovers more admire by far
> Thy naked beauties – give me a cigar!'

Although both the Romantic movement and the Peninsular War had paved the way for the reintroduction of smoking to Britain's upper echelons, it was far from being considered acceptable in society at large. If done at all, smoking was considered an outdoor activity and did not yet feature at the overcrowded, senseless 'rallies', 'routs' and other social gatherings that characterized the age. Genuine social acceptance of smoking and the cigar only occurred upon their adoption by the British cavalry. The convenience of cigars recommended itself to cavalry officers, who found it hard to manipulate snuff boxes when in the saddle. A cigar, however, could be wielded at the same time as a lance or a sabre, and by happy coincidence, was well suited to the bushy facial hair favoured by the cavalry officers of the period.

After Britain's martial successes on the continent and the thrill and pomp of the London celebrations that followed, it became all the rage to dress up and fight the French. Dandies rushed to purchase commissions in Britain's cavalry regiments where they equipped themselves with beautiful uniforms and blood horses. They developed new models of gallantry and of wit, and learned to smoke cigars, all three of which skills are demonstrated in the following extract from *Vanity Fair*, where a dandy turned cavalry officer is addressing Becky Sharp, the book's vivacious, if impressionable heroine:

'You don't mind my cigar, do you, Miss Sharp?' Miss Sharp loved the smell of a cigar out of doors beyond everything in the world – and she just tasted one too in the prettiest way possible, gave a little puff, and a little scream, and a little giggle, and restored the delicacy to the Captain, who twirled his moustache, and straightaway puffed it into a blaze that glowed quite red in the dark plantation, and swore 'Jove – aw – Gad – aw – it's the finest seegaw I ever smoked in the world – aw' for his intellect and his conversation were alike brilliant and becoming to a heavy young dragoon.

British cavalry officers soon had the opportunity to prove their martial, as well as social, skills when Napoleon slipped away from exile in Elba and his veterans rallied to him for a second attempt at European domination. The allies hastily reconvened their armies and the opposing factions converged on the Belgian village of Waterloo.

The apparent obsession with appearance in British society disguised its contempt for personal danger. The British cavalry was world renowned for its bravery, beauty and senselessness. Tactics consisted of a single manoeuvre – the charge, which was executed with the same commitment as employed in the pursuit of a fox. Their debut engagement against the French at Quatre Bras, on the eve of the battle of Waterloo, impressed a veteran of the Peninsular War: 'It did one's heart good to see how cordially the Lifeguards went at their work; they had no idea of anything but straightforward fighting, and sent their opponents flying in all directions.' Foot soldiers were amazed to see British cavalrymen who had been unhorsed retire from the field to change their clothes – they did not wish to be seen fighting in dirty uniforms.

The cavalry's reputation was confirmed and enhanced two days later at the battle of Waterloo, where the Union Brigade counter-attacked a body of French cavalry on impulse, then pursued them through their own guns and several lines of infantry, coming to within 100 yards of Napoleon, before his lancers turned their exhausted horses. The remnants of the Union Brigade that returned to British lines retired to a forest at the rear to regroup, where they were surprised to discover thousands of Belgian troops, who had fled there upon the French's first advance: 'I peeped into the skirts of the forest, and truly felt astonished; entire companies seemed there, with regularly piled arms, fires blazing under cooking kettles, while the men lay about smoking as coolly as if no enemy were within a day's march.'

Whether charging the French or smoking in the woods, the allies were victorious and Napoleon was exiled to the British island of St Helena in the South Atlantic, where he ended his days, his snuff supplied gratis by his conquerors. Upon their return to London the heroes of Waterloo were fêted. Viscount Wellington was created Duke of Wellington, and, as in the Elizabethan age, triumph overseas was celebrated with tobacco smoke in Britain.

Smoking's social revival was rapid and seemingly inexorable. In 1800, England imported 26 pounds of cigars. In

1830, it imported 250,000 pounds, including its first direct shipment from Cuba. This increase was assisted by a reduction in duty on cigars – almost the only occasion this has occurred in Britain. Smoking was not, however, without its detractors. In the half century that snuff had ruled society people had not only become much cleaner but also unused to the smell of burning leaves indoors. Some of the old order found smoking offensive and for the first time in Britain the segregation of smokers took place.

A chamber in the House of Commons was designated its 'smoking room', where those who wished to indulge were expected to retire. The atmosphere in this room is described by Macaulay, in 1831: 'I am writing here at eleven at night, in this filthiest of all filthy atmospheres . . . with the smell of tobacco in my nostrils . . . Reject not my letter, though it is redolent of cigars and genuine pigtail.' The introduction of smoking rooms illustrates that the revived habit was no longer perceived as a panacea – a universal medicine – and having lost this lustre, the fumes smoking generated could not be justified to non-smokers.

The exacting standards of personal hygiene Beau Brummell had pioneered also engendered resistance to the smell of tobacco smoke on smokers' hair and clothing. As a result, smoking jackets and smoking caps were developed, these latter modelled on the Moroccan fez. Smokers would don smoking garb whenever they wished to indulge, so that only a portion of their wardrobe would be tainted with the scent of tobacco smoke. The French still wear a descendant of the Regency smoking jacket – 'Le Smoking' – to their weddings. Beau himself did not witness the revival of smoking. The paragon among snuffers was forced into exile in Calais by his creditors whence he wrote begging letters to his old London suppliers, assuring them that there was 'not a good pinch of snuff to be had throughout France'.

The explosive growth of cigar smoking led to the establishment of a number of tobacconists dedicated to their importation. Foremost among these was Lambert and Butler of Drury Lane which began life as an importer of cigars and pipe tobacco. Its co-founder, Charles Lambert, 'was considered the best judge in London of an Havana cigar'. However, some of the more conservative establishments in London, particularly gentlemen's clubs, resisted the advance of smoking, and dedicated venues, named 'divans', sprung up around the West End to cater to smokers. Divans were modelled on the Ottoman sanctuaries for hookah smokers, and were fitted up 'in a style of Asiatic splendour and comfort, that produces to the uncultivated eye a very novel and pleasing effect; while upon a closer examination, the other senses are no less delighted'.

Twenty years after victory at Waterloo, not only was the cigar well established in polite circles, but the pipe had been revived. Britons returning from India had introduced the subcontinental version of the narghile – the hookah – to society, where its use was common enough to merit the inclusion of a hookah smoker in *Vanity Fair*. However, in this incarnation, a pipe was not a young man's tool, for that role had been taken by the cigar. Instead, pipe smoking was associated with age and statescraft. The example set by Marshal Blucher of Prussia on his visit to London was followed by British politicians, whom, with the decline in the power of and respect for monarchies, had become social figureheads. A pipe in the mouth of a politician was considered to confer gravitas. In Thackeray's words: 'The pipe draws wisdom from the lips of the philosopher, and shuts up the mouth of the foolish; it generates a style of conversation, contemplative, thoughtful, benevolent and unaffected.' A final consequence of the smoking revival was its readoption by English writers, so that tobacco resumed its honourable association with the language's creative geniuses. Many others followed Sir Walter Scott, William Wordsworth, Coleridge, De Quincey, Byron and Shelley in seeking inspiration in the weed's smoke, some, such as Charles Lamb, regarding it an indispensable part of their identity:

May my last breath
be drawn through a pipe
and exhaled in a jest.

Like the British, the Prussians had carried the cigar home
with them after their victory at the battle of Waterloo. The
cigar had hitherto been a novelty in Prussia. Less than ten
years previously, its appearance in the kingdom had been
noted in simple and enthusiastic terms: 'we must mention
here a new type of smoking, namely cigaros; these are
tobacco leaves rolled into hollow cylinders of the thickness
of a finger. One end is lighted and the other put in the
mouth. This is how they are smoked. This method, which
is used in Spanish America instead of a pipe, is beginning
to be common in our parts also; but whether it makes
tobacco taste better or no is hard to determine, for it is –
a mere question of taste.' However, cigars did not enjoy the
same level of acceptance in Prussia as in Britain. Some of
King Friedrich the Great's prohibitions against smoking out
of doors were still in existence and were enforced by zealous
policemen, especially in Berlin. They had been relaxed,
briefly, in 1809, when the French captured the city and
smoked in the streets to tease the vanquished Berliners. Upon
the expulsion of the French, the Berlin police were quick to
repair this breach in public order. Even in a state of war,
the citizens could not be allowed to relax. As a police officer
reported in incredulous tones: 'it is indescribable how far
this unseemly practice of smoking has gone; last evening,
three young men were sitting on the promenade each
smoking a long pipe; I sent Constable Schultz to stop them,
but they made off as soon as they saw him'. Even the
returning heroes of Waterloo were subjected to these same
school rules. In 1830, the Prussian government enacted a
law which prescribed that cigars should be fitted with a
wire-mesh fire guard designed to prevent sparks setting fire
to ladies' dresses.

However, tobacco use in Prussia was not confined to a
game of cat and mouse between smokers and the police. It

had become a matter of interest to scientists, who were curious to discover why people had started smoking in such numbers. For nearly a century, there had been little movement in the debate over the question 'Why smoke?' As religious opposition had declined, and after Napoleon had altered the balance of power in Europe, controversy surrounding smoking had subsided. No new reasons for using tobacco had been formulated, and some of the old justifications had been forgotten. Few people took tobacco for the good of their health – most defended its use on the grounds of companionship, fashion or stimulation.

Scientists were assisted in satisfying their curiosity about tobacco by contemporary advances in medical science, which had rendered the Galenic concept of humours redundant, and by the new science of chemistry whose techniques of refinement, isolation and identification had originally been developed in the laboratories of alchemists. According to chemists, the term 'elements' encompassed a greater quantity of substances than earth, air, fire and water, and the properties of materials could be determined with better precision than 'hot' and 'wet'. This greater sophistication of analytical method enabled mankind to increase its understanding of the invisible processes of creation and destruction, and what it was in tobacco that people craved.

Chemists began to focus their research on the isolation of the active physiological ingredients of plants. Why did some vegetables have specific and predictable effects on the functions of the human body? Chemicals were obviously at work, but which ones? In 1803 Friedrich Wilhelm Serturner isolated a substance he named 'morphine', after the Roman god of pleasant sleep and sweet dreams. This essence was the purified spirit of the opium poppy – its active ingredient. The great Belgian chemist, Joseph Gay Lussac, whose fireside pipedream of a serpent clutching its tail unlocked the secrets of organic chemistry, proposed that all similar plant extracts, if discovered, should have the suffix '-ine' added to their names. Scientists now raced each other to discover other members of this class. In 1809, Vauquelin extracted a 'potent, volatile and colourless substance' from tobacco which he named *essence de tabac*, as if it were an after-shave. *Essence de tabac* was, however, impure and the next '-ines' to be discovered were strychnine, quinine and, in 1820, caffeine. The triumph of isolating the tobacco '-ine' was left to two Heidelberg students, Ludwig Reimann and Wilhelm Heinrich, who succeeded in 1828, naming their discovery 'nicotine' – another posthumous honour for Catherine de' Medici's trusty court physician.

The discovery of the weed's narcotic ingredient launched an avalanche of further research. Scientific journals reported the results of various experiments with nicotine. It was demonstrated to be effective against a variety of afflictions, including disorders of the nervous system, haemorrhoids (via a tobacco enema) and several fatal diseases including malaria and tetanus. Other therapeutic properties attributed to nicotine included its effectiveness as an antidote to strychnine and various other poisons, such as snake venom. In these respects nineteenth-century science supported the age-old claims made for tobacco by Amerindians.

Further properties and applications nicotine might possess were also investigated. Nicotine's scientific trials coincided with the adoption of systematic animal testing. While animals had been used for centuries to demonstrate the efficacy of poisons, such experiments were usually conceived of as mere spectacles, and animals were not bred for the specific purpose of determining safe dosage levels of particular substances. Scientists annihilated entire kennels of dogs with nicotine injections, leading them to conclude it was a powerful poison. It was determined that the quantity of nicotine contained in an ordinary cigar – if it were extracted and injected – would prove lethal for two adult humans.

Surprisingly, this discovery did not inspire the researchers to dissuade people from tobacco consumption. Most of the other '-ines' they had discovered were also poisonous in the right quantities, some such as caffeine and strychnine spectacularly so. Such further investigative work as was carried out centred on the molecular structure of nicotine which it was hoped might be synthesized artificially for use as a drug to combat various ailments, thus converting tobacco from a folk remedy into a scientific cure.

9

The American Preference

*The Americans explore their new country –
tobacco in the wild west – the habit of
chewing tobacco and spitting – the story of
Carmen and the birth of the cigarette – the
fight for the right to smoke in Prussia –
the anti-smoking revolt in Italy*

While chemists were investigating the chemical causes of
mankind's affair with tobacco, some of the ritual reasons
for its consumption came under investigation in the United
States of America. Napoleon was the cause of this scrutiny.
In addition to financing his martial adventures with tobacco
taxes, he had raised funds by selling the chunk of America
France owned to the United States in 1803. This transaction,
known as the 'Louisiana Purchase', doubled the size of the
United States. Its third president, the former tobacco farmer
Thomas Jefferson, wanted to establish what he had bought,
as the French vendors had left him in some doubt. When
Napoleon's minister Talleyrand (whom Napoleon described
as 'filth in silk stockings') had been asked to point out on
a map exactly what the United States had paid for, he replied
he could give them 'no direction'. Jefferson was also curious
to know what lay between his new purchase and the Pacific
Ocean. He dispatched two surveyors, Meriwether Lewis and
William Clark, on an epic journey across America. Their
mission was to follow the Missouri to its source, to find

what rivers emptied west into the Pacific and discover if a navigable transcontinental route existed.

It was not then known if the interior was composed of mountains or plains, of arid deserts or wetlands. It was believed to harbour tribes of Indians, but their numbers and disposition were a mystery. Lewis and Clark's expedition was ethnographical as well as geographical. They were instructed to go in peace and to make friends with whatever tribes they encountered. To this end, they were equipped with tobacco – a continental symbol of friendship and its currency. Tobacco served as their passport to the unknown regions of their homeland. Whenever they passed into the territory of a new tribe, the first preliminary to goodwill was to sit down and smoke with them. Lewis and Clark's mission was pursued jointly to the shores of the Pacific Ocean, and separately on the return east, when they took different routes home. Both were back in Washington by 1806, and when the results of their surveys were published a wave of enthusiasm swept through the citizens of the United States as they contemplated a migration west.

They were motivated, in part, by similar impulses to their English forebears – they believed they had been selected by God, or destiny, to civilize heathens and to cultivate wildernesses. They were exhorted from the pulpit with soaring eulogies that hinted at the glories that awaited them: 'The far reaching, the boundless future will be the era of American Greatness. In its magnificent domain of space and time, the nation of many nations is destined to manifest to mankind the excellence of divine principles: to establish on earth the noblest temple ever dedicated to the worship of the Most High.' The citizens of the United States were also encouraged to embrace their manifest destiny in more homely language, whose identical aim was to send them racing in the direction of the sunset: 'Land enough! Land enough! Make way, I say, for the young American Buffalo – he has not yet got land enough! He wants more land as his cool shelter in summer – he wants more land for his beautiful pasture grounds.'

Migration proceeded cautiously at first as new trails were

forged – due west, towards California, and north-west, via
the Oregon trail. The distances across the continent were
immense. Nothing like them had been attempted overland
in Europe since the days of Tamerlane the Great. Even
Napoleon's disastrous march on Moscow would have taken
him no more than halfway to California. Trails *per se* were
neither roads nor paths. Migrants found their way by
following footprints and the skeletons of men and animals.
They travelled by horse, on foot and in convoys of oxen
carts, driving their cattle alongside. For the early decades of
the nineteenth century the 'wild west' consisted of Cleveland,
Ohio, and Pittsburgh. But every day, explorers were creating
new routes, laying trails that migrants might follow. In 1820,
a trail was opened to Santa Fe in Hispanic California, where
adventurers were amazed to find ladies of that city smoking
seegaritos. Female migrants did not, as a rule, smoke. It was
considered to be a man's habit, especially by the war-like
tribes through whose territory the migrants passed.

The migrants' interaction with the Indian tribes did not
follow the same peaceful course established by Lewis and
Clark. The settlers, indeed most Americans, believed that
Indians were dangerous savages who should be eradicated
from the land. By the time the trickle of migrants had turned
into a flood, the country had a president, General Andrew
Jackson, who was of the same view. Jackson, who had killed
redcoats, senators and Indians in his time, reckoned the
Indians the most dangerous of these opponents and, since
they had no other merits, deserving of extermination. He
had, however, picked up one Indian habit during his years
of campaigning. Jackson smoked 'a great Powhatan bowl
with a long stem' puffing out until the room was 'so obfus-
cated that one could hardly breathe'. He also had a very
Indian approach to relaxation. As he commented during his
tenure of the newly built White House: 'Mrs Jackson & I
go to no parties [but remain] at home smoking our pipes.'

The western exploration of America created a second
opportunity to observe native Indian tobacco use as it had
existed since humans had arrived on the continent. Since

tobacco was the migrants' key to communication with the tribes they despised, and smoking was no longer strange to them, they examined the Indians' habit minutely with expert eyes. These early nineteenth-century observations were to prove influential on smoking habits in Europe, in particular when documented with a Romantic pen. One such observer, Francis Parkman, a Harvard graduate and admirer of Lord Byron, who conducted a six month hunting expedition from St Louis to the Black Mountains in 1846, described life and smoking among the Dakota Indians, whom he considered, whilst enchanting, to be 'thorough savages. Neither their manners nor their ideas were in the slightest degree modified by contact with civilization. They knew nothing of the power and real character of the white men, and their children would scream in terror at the sight of me.' This was one of the last pure contacts between Indians and white men. In the years that followed, the Indians surrendered their culture to the settlers, whose only reciprocation was to admire their pipe rituals. Parkman observed the omnipresence of recreational smoking among both the tribes and the frontiersmen who traded furs with them. Whenever men considered themselves to be at rest, whether they were merely pausing on the trail, or feasting when encamped at the summer hunting grounds, a pipe would be lit and passed from hand to hand.

Parkman also recorded the official functions of tobacco, including its ability to serve as a sign of both peace and of war. He describes one incident in which the Snake tribe had inadvertently murdered the son of a Sioux chieftain whose scalp they sent to his father together with a parcel of tobacco which signified that they acknowledged their mistake and wished to stay at peace. In response, the Sioux chief, known as the 'Whirlwind', sent messengers with tobacco to all the other Sioux tribes, in this case the tobacco serving as a confirmation of kinship and as a summons to war against the Snakes. Westerners soon learned and were careful to observe pipe etiquette, knowing whether to smoke or to abstain by the direction in which the pipe was passed: 'A large circle of warriors were again seated in the center of the village, but this time I did not venture to join them, because I could see that the pipe, contrary to the usual order, was passing from the left hand to the right around the circle, a sure sign that a "medicine-smoke" of reconciliation was going forward, and that a white man would be an unwelcome intruder.'

It was not only the ubiquity of smoking that attracted the attention of migrants and visitors west, but the veneration surrounding tobacco pipes. Western culture considered a pipe to be a practical, not a ceremonial object, and the possibility that it could be more intrigued them. It was like attaching equal importance to a wine bottle as to its contents. Indian pipes became incorporated into Western fable, perhaps most famously in *The Song of Hiawatha*, Henry Longfellow's epic poem. In some cases Indian legends surrounding tobacco pipes turned out to be true. George Catlin, a landscape and portraiture artist who spent eight years travelling amongst the plains tribes, was the first white man to see the celebrated pipe mine of Indian legend:

we found the far-famed quarry or fountain of the Red Pipe, which is truly an anomaly in nature. The principal and most striking feature of this place is a perpendicular wall of close-grained, compact quartz, of twenty-five and thirty feet in elevation, running nearly north and south, with its face to the west, exhibiting a front of nearly two miles in length, when it disappears at both ends, by running under the prairie . . . At the base of this wall there is a level prairie, of half a mile in width, running parallel to it, in any, and in all parts of which, the Indians procure the red stone for their pipes . . . From the very numerous marks of ancient and modern diggings or excavations, it would appear that this place has been for many centuries resorted to for the red stone; and from the great number of graves and remains of ancient fortifications in the vicinity, it would seem, as well as from their actual traditions, that the Indian tribes have long held this place in high superstitious estimation; and also that it has been the resort of different tribes, who have made their regular pilgrimages here to renew their pipes.

'Catlinite', as the fine grained silicate stone from the quarry was named, had been found in pipes as far away as Quebec.

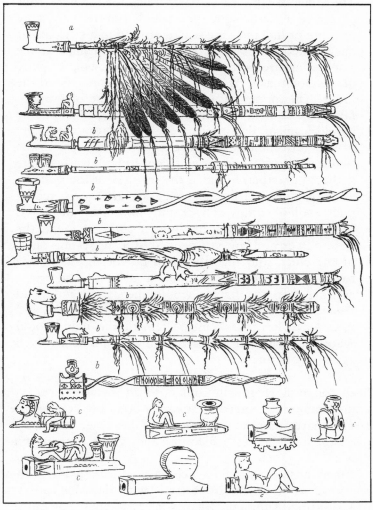

George Catlin – Indian tobacco pipes

The pipe became a central feature in American descriptions of their Indians and was indelibly associated with the red-men in their minds. Here is an example by Parkman, drawn from the flesh, of a night in a Sioux lodge:

> The lodge of my host Kongra-Tonga, or the Big Crow, presented a picturesque spectacle that evening. A score or more of Indians were seated around in a circle, their dark naked forms just visible by the dull light of the smoldering fire in the center, the pipe glowing brightly in the gloom as it passed from hand to hand round the lodge. Then a squaw would drop a piece of buffalo-fat on the dull embers. Instantly a bright glancing flame would leap up, darting its clear light to the very apex of the tall conical structure, where the tops of the slender poles that supported its covering of leather were gathered together. It gilded the features of the Indians, as with animated gestures they sat around it, telling their endless stories of war and hunting. It displayed rude garments of skins that hung around the lodge; the bow, quiver, and lance suspended over the resting-place of the chief, and the rifles and powder-horns of the two white guests. For a moment all would be bright as day; then the flames would die away, and fitful flashes from the embers would illumine the lodge, and then leave it in darkness. Then all the light would wholly fade, and the lodge and all within it be involved again in obscurity.

Such romantic representations outlived their subjects – the real Indians were forced westwards, bound to treaties their counterparts had no intentions of keeping, infected by diseases, introduced to alcohol, and, from time to time, massacred.

Americans also came into contact with the west of their continent via their maritime trade with the Mexican state of California. Although twice as far away by sea as Europe, it was four or five times more lucrative as a market, and enterprising Yankee ship-owners carried cloth, clothing, crockery, ironware and trinkets around the southern tip of the Americas to trade for silver and hides on the Pacific coast. Owing to the total absence of industry in California, the hides carried back to the east coast were often made into shoes to be shipped back round Cape Horn on the next voyage.

Mariners engaged in the Californian trade encountered a different tobacco habit to the frontiersmen. Hispanic Californians smoked cigars, which the sailors came to prefer to their pipes. They took both pipes and cigars onwards across the Pacific, and were the first to bring a regular supply of tobacco to the Sandwich Islands (Hawaii), with which they traded. The Sandwich Islanders, whose ancestors had killed and eaten parts of Captain Cook, evolved a highly individual method of smoking, more reminiscent of that of the plains tribes than the white man from whose hands they had received tobacco:

> They smoke a great deal, though not much at a time, using pipes with large bowls and very short stems, or no stems at all. These they light, and, putting them to their mouths, take a long draught, getting their mouths as full as they can hold of smoke, and their cheeks distended, and then let it slowly out through their mouths and nostrils. The pipe is then passed to others, who draw in the same manner, one pipeful serving for half a dozen. They never take short, continuous draughts, like Europeans, but one of these 'Oahu puffs' as sailors call them, serves for an hour or two, until someone else lights his pipe, and it is passed around in the same manner.

At the same time as they were observing and documenting native pipe cultures, Americans had become enamoured of the Spanish cigar, in particular on the eastern seaboard, which enjoyed a flourishing trade with Cuba, legal after 1817, when the Spanish crown had allowed its colony to conduct business with other countries. The Decreto Real of 1817 relaxed some of the Spanish monopoly controls on the production and sale of tobacco, in order to stimulate the Cuban industry. These steps had been taken in the hope of preserving Cuba's loyalty to Spain, which lost most of its American empire in the first quarter of the nineteenth century. Gone were the silver mines of Peru, Mexico had finally broken free after years of lawlessness, taking Florida and California with it. Cuba, the 'Pearl of the Antilles' had to be humoured lest it ran away as well.

Cuban tobacco production flourished with the arrival of free trade. Havana cigars found a ready market in Europe, where it had been discovered that cigars kept their freshness better on a long sea voyage than bulk tobacco, resulting in a preference for Cuban cigars over cigars rolled in Spain with the same weed. However, Cuba's principal market was the United States, whose citizens had had a taste for its cigars since General 'Abe' Putnam had participated in the British sack of Cuba in 1762. Putnam had loaded three donkeys with Havana cigars as his share of the plunder, which he sold singly to customers of a tavern he owned in Connecticut. Havana cigars had been an item of trade ever since, and Cuban seed was imported to Connecticut which started to produce cigars of its own.

American demand for Cuban cigars created a boom in the Spanish colony. More and more land was planted out to tobacco, and cigar factories dominated Havana's skyline. Already a mystique had attached itself to Cuban cigars. Americans considered them superior in manufacture and flavour to their own creations, including the Conestoga cigars or 'stogies' which were favourites among migrants west. Cubans exploited this preference by branding their products. The first grand marque of Cuban cigars, Ramon Allones, was created in 1827. By 1845, tobacco had replaced sugar as Cuba's principal export. Within a further ten years, Cuba had 9500 tobacco plantations, nearly 2000 cigar manufacturers and the trade employed in excess of 15,000 people. By this time, the process of rolling cigars had arrived at the

standard in use today. This involves up to three types of
tobacco: the filler, comprised of chopped or folded tobacco
leaves forming the 'meat' of the cigar, surrounded by a single,
pliant leaf called the binder, around which is rolled the
wrapper, a leaf selected for its colour and flavour. A cigar
should have proportion, in that its size and shape are in
harmony with its flavour and strength. This proportion is
known as a cigar's *vitola* – a kind of ideal balance of form
and function that interacts with the smoker. As Don Fernando
Ortiz, author of the seminal 'Cuban Counterpoint of Sugar
and Tobacco' observed 'a cigar's vitola is part and parcel of
the smoker's own vitola'.

Despite the ubiquity of smoking in the wild west, and
rising demand for Cuban cigars along the east coast, the principal tobacco habit in the United States was smokeless. For
every man who lit a cigar or inhaled from the sacred calumet,
ten more took their tobacco raw. To many Americans, the
question 'Why smoke?' would have been an enigma. The
weed was not for smoking, but chewing, a practice which
had the side effect of generating copious quantities of tinted
saliva which the chewer would eject from time to time by
expectoration. Chewing was an ugly habit to exhibit to visitors, who coincidentally had begun to arrive from the Old
World. In the same decades as the Americans were exploring
their own country and enlarging it through theft, war, treaty
and purchase, European visitors to the United States were as
eager to observe the habits and customs of Americans as the
objects of their attention were to scrutinise the Indians.
Among such visitors was the rising star of English fiction,
Charles Dickens. Dickens was a man of his age and passed
judgement on whatever scene or person he described, which
vice made him a favourite among the subjects of the newly
crowned Queen Victoria, who liked a moral on every page.

Dickens travelled throughout the United States between
January and June of 1842. The Americans expected their
visitor to be enchanted – he had admired the Republic from
afar, and it was anticipated that affection could only turn
to love upon encounter. Unfortunately, the prevailing

naked Andalusians, who also smoked. Seville, its women and its cigars had also made a deep impression on the Frenchmen who had occupied it earlier in the century, and it became a favourite destination of French literati in the 1830s, who celebrated a new use for tobacco – as a sexual ambassador – on their return to France. The first of these to venture south and document the phenomenon was Prosper Mérimée, a prominent figure in the French Romantic movement. Seville's tobacco factory gave him the material for his most famous work – *Carmen* – the tale of a gypsy temptress who breaks hearts and steals watches. This is how Carmen dressed for work at the Fabrica:

> She was wearing a very short red skirt, beneath which you could see her white silk stockings with holes in them and dainty red morocco leather shoes fastened with flame coloured ribbons. Her mantilla was parted so as to reveal her shoulders and a big bunch of acacia flowers which she had in the front of her blouse. She had another acacia bloom in one corner of her mouth, and she moved forward swaying her hips like some filly out of the Cordoba stud. In my part of the world everyone would have crossed themselves at the sight of a woman dressed like that; but there in Seville everyone paid her some risqué compliment on her appearance.

Carmen would have removed everything but the stockings and flowers when she sat down to roll cigars.

Interestingly, Carmen the fictional heroine may have been drawn from life. The *cigarreras* were as wild as their reputation and the Fabrica's records from the time of Mérimée document the expulsion of one Maria del Carmen Garcia, a dark haired black-eyed 15-year-old firebrand who had been disciplined numerous times, and who was finally expelled from the factory after she attacked a workmate with a pair of tobacco shears.

The French appreciated the idea of beautiful Spanish girls making things to smoke, and other curious writers followed

Mérimée's trek south to observe the process. Here are Pierre Louÿs's notes on the Fabrica's employees:

> this immense harem of four thousand eight hundred women is as free in speech as in words . . . they showed no reserve in profiting by the tolerance which permits them undress as much as they like in the insupportable atmosphere they live in from June to September . . . almost all worked stark naked to the waist with a simple linen petticoat unfastened round it and sometimes turned up as far as the middle of the thighs . . . and they presented an extraordinary mixture . . . there was everything in this naked crowd, virgins excepted, probably.

The girls themselves did not smoke cigars but *papelotes* – the shredded tobacco wrapped in paper that was the favourite of Spain's poor. *Papelotes* were soon in vogue amongst visiting French writers, being a necessary part of their armoury of seduction. When these men returned to Paris, they carried *papelotes* with them where the device was renamed. Its new appellation is now the most commonly used French word on the planet: 'cigarette'. Gautier had the honour of its baptism. He too had visited Seville and commentated on the visible charms of its tobacco workers. One of the heroes of his *Les jeunes-France* (1833) is presented to the reader 'nonchalantly smoking a little Spanish cigarette'. By 1840, these curiosities had become common on Parisian streets, and had developed an identity. Theodore Burette, a smoking commentator, observed in his *La physiologie du fumeur*: '*La cigarette est gentille, vive, animée: elle a quelque-chose de piquant dans ses allures. C'est la grisette* ["the babe"] *des fumeurs.*' Interestingly, Burette sexed the cigarette as a woman. It seems that even on its baptism, the cigarette was fated to be associated with eroticism and urban sexuality. Charles Baudelaire, master of lust and melancholy, also commented on the new addition to the smoker's arsenal, describing the prostitutes in the quartier around the Church of Our Lady in the Ninth Arrondissement smoking cigarettes 'to kill time', which he recorded in the *The Salons of 1848*. Tobacco from its earliest days had been used to encourage the acquaintance and bonding between fellow men. The flirtatious cigarette workers in *Carmen* and Baudelaire's languorous tarts were amongst the first to display its power to create a similar bond between men and women.

The feminine cigarette was a revolution, even in amoral France, where all other smoking devices carried the usual associations of masculinity, friendship and thought. Tobacco was a magician, not a witch, and even the syphilitic hashish addict Baudelaire expressed his debt to the weed in conventional form. Here is his pipe talking to us, and incidentally explaining why most nineteenth-century Frenchmen smoked:

> I am a writer's pipe. One look at me,
> and the coffee colour of my Kaffir face
> will tell you I am not the only slave:
> my master is addicted to his vice.
>
> Every so often he is overcome
> by some despair or other, whereupon
> tobacco clouds pour out of me as if
> the stove were kindled and the pot put on.
>
> I wrap his soul in mine and cradle it
> within a blue and fluctuating thread
> that rises out of my rekindled lips
>
> from the glowing coal that brews a secret spell.
> He smokes his pipe, allaying heart and mind,
> and for tonight all injuries are healed.

As cigarettes progressed from being a foreign novelty to a common Parisian habit, they came to the attention of France's state tobacco monopoly, SEITA, which had survived, or been reinstated, through the course of a revolution, a republic and a tyranny. Its then beneficiary, Louis Napoleon III, was happy to see Frenchmen smoking. 'This vice brings in one hundred million francs in taxes every year,' he remarked when asked to take action against the habit. 'I will certainly forbid it at once – as soon as you can name a virtue that brings in as much revenue.' In accordance with the wishes of its ruler, the French state went into the cigarette business. In 1845, the first year of cigarette production, it sold 6 million. SEITA was perhaps unique among tobacco monopolies of the age in that it sought to encourage French citizens to smoke, anticipating changes in their tastes with alterations to its products. For instance, it was noticed that cigarette smokers preferred Virginian weed and the ingredients of SEITA's cigarettes were altered accordingly. Not all of the monopoly's clients for cigarettes were Parisian prostitutes. Napoleon III became a fervent cigarette smoker, particularly on battlefields, which he was obliged to attend when France engaged the German state of Prussia in another contest for dominance in continental Europe.

Tobacco had continued to follow a hard road in Prussia. Despite general acceptance of smoking in the rest of Europe, it was still prohibited on the streets of Berlin. The ban had been relaxed briefly in 1831, during an outbreak of cholera, when smoking in the public streets and squares was permitted, 'so as not to deprive anyone of any possible protection against infection'. Berliners had to wait another six years for the next outbreak of cholera when the ban was again relaxed and they could take their pipes and cigars into the streets. This state of affairs continued into the 1840s as is evinced by the petition of a painter named Otto Gennerisch to the Prussian king: 'In filial reliance on the paternal ruler who has given us freedom of thought, and thereby let us know that he would like to see the barriers between sovereign and subjects removed, as rendered necessary by this age of progress, I venture to present a petition . . . in Italy smoking is allowed everywhere; we now ask your majesty that we Berlin smokers may claim a similar compliment.' It is fascinating that Gennerisch prefaced his petition with an expression of gratitude for being allowed 'freedom of thought', a liberty only recently granted by the Prussian king to his subjects, prior to which, presumably, they had been acting 'under orders'.

Berliners had to wait until the revolution of 1848 for their deliverance from anti-smoking regulations. The right to smoke in public was a specific demand of the successful revolutionaries. Governments seldom learn the lesson that however trivial a protest may appear to be, if the grievance is universal, they ignore it at their peril. All of Europe was in turmoil at the same time, and the right to smoke featured as a cause in another of the struggles for independence that occurred that year.

A true mirror image of the Berliners' fight for the right to smoke in public took place in the Italian states of Lombardy, Venetia and Piedmont, then under Austrian control, where the right not to smoke was advanced by Italian patriots in the interests of securing further, more fundamental rights. Tobacco was distributed in these states under the Austrian state monopoly and its products were selected as a symbol of oppression. Italians were exhorted to demonstrate heroic self-restraint and boycott the weed: 'Franklin's fellow citizens gave up their tea; do you imitate them and refuse Austrian tobacco. This is no vain attempt, but a duty, a master stroke, and a sign of union and unity. We must make some sacrifice . . . who will dare to assert that this habit has become a necessity to Italians? A nation that wishes to rise must love its country and help her as best it can.'

The greatest no-smoking protest of the nineteenth century began in Milan on New Year's Day, 1848. Cigars were knocked from the mouths of anyone seen smoking on the streets and even soldiers assaulted. The Austrian troops retaliated by forcing their way into restaurants in groups of twenty or more, all with cigars in their mouths. Some detachments were ordered by their officers to smoke whenever on patrol. The protest spread to other cities beyond Milan, causing alarm in official quarters. Field Marshal Count

Radetsky, commander of the occupying forces stood firm:
'I will not recognise or tolerate any secret society that insults
and attacks peaceful smokers in the streets,' he wrote to
Archduke Rainer in Vienna and declared his intention of
proclaiming martial law. The 'cigar disturbance' as it was
described spread to Padua and Venice where Metternich, the
Austrian chancellor and master statesman of his age, of
whom Napoleon had observed 'he lies all the time and that
is too much', was advised: 'This cigar business is in itself,
as you rightly say, childish; but it is an attempt of a faction
to stir up the mob, and in that aspect it is not so unimportant. It would be most unpleasant if we had to use force
of arms, at a time when everything is represented in the
most odious light in the papers.'

The cigar revolution progressed to open revolt in
Lombardy and Venetia, and to general war in Piedmont.
Austrians were forced to evacuate their troops from Milan
and surrender the city where they had previously smoked
with the impunity of oppressors to the Italian patriots. While
the uprising was suppressed within a few months, it was the
first blow struck for a free and united Italy, and the first
example of a body of people asserting their right not to
smoke.

Tobacco was also to feature in a conflict in the United
States. The tobacco growing states were slave states, and
this form of human bondage, already considered despicable
in Europe, had become a fundamental point of difference
amongst the citizens of the United States. America had only
hitherto avoided a civil war through the efforts of the single
American Dickens had wholeheartedly admired in the course
of his visit: Henry Clay (1777–1852), the abolitionists' champion in the US Senate. Clay neither slaved nor spat, and his
vigorous speeches advocating universal human rights
endeared him to philosophers and revolutionaries around
the world. After a visit to Cuba in 1850 a cigar brand was
named after him as a gesture of respect. Clay, a giant
Kentuckian known for his womanizing and table dancing
exploits, was nicknamed the 'Great Compromiser' on

account of his efforts as a mediator to preserve union between the pro-slavery and abolitionist states. Within a few years of his death war had commenced between America's northern and southern states, which culminated in a victory for the north and the abolition of slavery.

Change was also at hand for the tobacco-eating habit Dickens had detested. The substance which would one day supersede chawing and become the most widely consumed variety of vegetable on earth had been discovered prior to his visit, on a farm in Caswell County, North Carolina. The soil in this region produced a lighter coloured tobacco leaf known as 'Piedmont', which burned with a smoother flavour. When, by chance in 1839, its curing process was altered by a slave named Stephen who used charcoal instead of logwood as fuel for the curing fires, the Piedmont leaf turned to gold. Of all the tobacco varieties yet cultivated, 'yallacure' as it was named, was the easiest to inhale. Western smokers seem to have forgotten inhaling while snuff was king, and in the nineteenth century a higher value was set upon sobriety than in previous smoking eras. A tobacco that could be inhaled without intoxicating the smoker was an important discovery. This compulsive substance was to become the bell-weather variety of cigarette tobacco.

10

A Class is Born

*The surprising world of Victorian smokers
– pipe rituals in British society – clysters
and the dangers of auto-intoxication –
smoking, women and children – smoking
and Empire*

Whilst the United States were fighting one another, cigarettes had made their debut in Britain. British soldiers had encountered them in various engagements abroad, commencing with the Peninsular War, but it was not until the 1850s that they were manufactured or sold in Britain. The British were introduced to the habit at home after the Crimean War, when a returning veteran, Robert Peacock Gloag, set up the first cigarette factory on British soil in 1856, which manufactured cigarettes in the Russian style: 'The tobacco used was Latakia dust, and the paper yellow tissue . . . the mouthpiece was of cane. The mode of manufacture was first to make the canes, into which the tobacco had been pressed. In order to keep the dust tobacco from escaping, the ends were turned in.' It seems, however, that the British were not yet ready for cigarettes and Gloag later turned to making cigars. Almost contemporaneously, the Bond Street manufacturer Philip Morris also began to turn out hand-made cigarettes.

The new cigarettes were entering a crowded market. The smoking revival in Britain had progressed to the extent that in the latter half of the nineteenth century excessive smoking was once again a national characteristic. The question 'Why smoke?' had metamorphosed into 'What do you smoke?' The cigar had established itself as the preferred implement of the rich. The clay pipe continued to enjoy pre-eminence among the rural poor. The pipe was also the badge of the intellectual and the statesman. In addition to consuming the fumes of their herbal friend, the British continued to use tobacco in other forms. Mighty quantities of snuff were snorted in the Shires and the new industrial towns of the north. The clyster, a device for delivering tobacco enemas, appeared in doctors' surgeries and private homes up and down the land.

The Victorian tobacco industry celebrated the diversity of its customers. Even the smallest manufacturer carried at least a dozen products. These included loose tobacco, in exactly the same forms as had been sold two centuries ago – shag, pigtail and twist, various branded blends of pipe tobacco and snuff, an in-house line of chewing tobacco, and from the 1860s onwards, a few varieties of hand-rolled cigarettes.

Victorian manufacturers believed that their customers wished to be treated as individuals as opposed to consumers, and catered to their whims accordingly. An exemplary practitioner of this business philosophy was Cope Brothers of Liverpool, a firm responsible for such forgotten gems as Court, Burgomaster and St George cigars. Copes' advertising emphasized the differences between its customers, and instead of seeking to convert them to a single product, aimed to demonstrate its ability to provide for all tastes. Copes published a series of popular wall hangings and calendars illustrating this united nation of tabagophiles. 'In pursuit of diva nicotina' is a typical example. It depicts a throng of happy smokers, many caricatures of great or notorious men, rushing towards their temptress, La Diva Nicotina.

The variety of habit confronting manufacturers was a result not only of local preferences, which were deeply ingrained, but also of social class. People were defined by what they smoked. Prince Albert, Queen Victoria's beloved,

In pursuit of diva nicotina

smoked cigars and installed a smoking room at Osborne on the Isle of Wight. Out of deference to his tabagaphobe spouse's feelings, this room was the only one in Osborne House with a solitary A instead of an entwined V&A over the door. Lord Alfred Tennyson, whose tendency to write poetry overruled his innate aristocratic leanings towards cigars, smoked a pipe, as did his fellow bohemian, the decorator William Morris. The class warrior, Karl Marx ('nothing can have value without being an object of utility'), consumed hundreds of cheap cigars while composing his political works. Lord Cardigan selected a cigar when he led the Charge of the Light Brigade. Oscar Wilde, the figurehead of decadence in that age, smoked cigarettes shamelessly.

Among this host of eminent Victorian smokers, one stands head and shoulders above the rest: the first man to prove to his fellow *Homo sapiens* that they were animals. Charles Darwin, the greatest scientist since Newton, was as fond of tobacco as his fellow immortal. Darwin was taught to smoke by the gauchos in Argentina, although his principal tobacco habit was snuffing. As his son recalled: 'He only smoked when resting, whereas snuff was a stimulant, and was taken during working hours.' In one of Darwin's early letters he speaks of having given up tobacco for a month, and describes himself as feeling 'most lethargic, stupid, and melancholy' in consequence.

Darwin had the good fortune, while developing his theory of evolution by natural selection, to witness the most significant social revolution that has ever occurred in Britain – the birth of the middle class. Conditions for this event were near perfect in the Victorian age. Money earned from manufacturing and overseas trade poured into the country, enriching a new type of person, many of whom intermarried. Soon they were numerous enough to merit their own position in the social strata. Simultaneously, posts for educated men became more plentiful and lucrative, swelling the professional caste around which the new grouping had been placed. The middle class, like a nest of industrious insects, built themselves an identity, of which pipe smoking was a part. As Darwin had argued, in the course of their social evolution they carried with them vestiges of their animal past.

Smoking gave a clarity to the class system through which a smoker could proceed as his wealth increased. How the habit was prosecuted defined the man. The middle class found its totem in the pipe. It was the perfect device for expressing both individuality and respectability. The greater importance accorded by the middle class to smoking led to design changes in pipes. The clays which had been used in more or less the same form since the introduction of smoking to England, were indelibly associated with manual labourers. Moreover, clays were highly perishable, which did not permit the middle-class smoker to bond with his instrument.

Such limitations, imposed by the frailty of the material itself, led to the introduction of 'meerschaum' pipe bowls from the continent. Meerschaum, German for 'sea foam', is a hydrated silicate of magnesia found mainly in Eskisehir in Turkey. Tobacco smoked from a meerschaum bowl was both cooler and sweeter in flavour than that burned in clay. Meerschaum also had the attractive property of changing hue from a corpse white to a honey brown over years of use. The colour of a meerschaum pipe was indicative of its owner's experience and smoking prowess, both desirable qualities to Victorians. Meerschaum owners were extravagantly protective of their pipes when these were in the process of being coloured. After use, the bowl was wiped with a silk handkerchief carried exclusively for that purpose, then stored in a tight fitting chamois pouch.

The potential of the care required for a meerschaum as a substitute for work appealed to certain members of the upper class. Cavalry officers adopted them, although it seems they were sometimes too impatient to carry out the colouring ritual themselves. Alfred Dunhill relates the tale of one enthusiast who enlisted a detachment of the Life Guards to smoke his pipe in an endless chain, 'swaddled carefully in soft flannel, and filled always with the best tobacco at the owner's expense, the pipe passed from mouth to mouth for the space of seven months. When it was finally unwrapped, it was coloured a rich, deep brown, pronounced perfect by connoisseurs of the art of pipe colouring.'

The meerschaum had a rival for the middle classes' affection – the briar wood pipe. Although wooden pipes had been attempted for centuries, their combustible properties shortened their lives, for none but the most diehard smokers could endure the taste of charcoal for long. The discovery of a type of wood suitably refractory to contain burning tobacco without itself catching fire may be attributed to Napoleon. Some years after the great snuffer's death, a Bonapartist pilgrim to the tyrant's birthplace in Ajaccio broke his pipe and requested a passing Corsican peasant to make him another. The peasant carved the pilgrim a pipe bowl from a local wood – *Erica arborea* or 'bruyère' – which functioned as effectively as his broken meerschaum. Inspired, the pilgrim sent samples of the same wood to his usual pipe-makers in St-Claude, near France's Jura mountains, who realized their potential and manufactured them commercially. Briar pipes, as these wonders were known in England, became available in the 1850s and were an immediate success.

Jealous, perhaps, of the colouring ceremonies enjoyed by meerschaum owners, the proprietors of briars soon found a way to discriminate amongst their possessions. This was based upon the straightness of the grain of wood used to make their pipes. The discovery of a method by which precedence might be established did much to assist the briar's popularity amongst Victorian smokers. Like the meerschaum, the briar had enhanced organoleptic qualities in comparison to the clay pipe, and both materials' cooler smoke enabled shorter stems. These were usually wooden, and tipped with a horn or amber mouthpiece. Amber was preferred on account of its expense, its elegance and its resistance to the teeth of ruminating smokers. Its values as a mouthpiece had first been established by the Persians, who believed amber purified whosoever touched it and had used it to tip their narghiles. Horn mouthpieces, by contrast, had a tendency to split in the mouths of pipe smokers with too tight a 'grip' as the habit of biting while smoking was known.

Together, the meerschaum and the briar enabled greater variety in both the shapes and the sizes of pipes which, given the idiosyncrasies of Victorian smokers, resulted in an eruption of new designs. The selection of a pipe became an essential ritual on the passage to manhood, and when blended pipe tobaccos were introduced the choice of an appropriate fuel for the owner's talisman also became a quest. Journals of the age ran regular tabagophile features, some of which, such as those by J. M. Barrie, author of *Peter Pan*, were collected into books. *My Lady Nicotine*, Barrie's offering, includes such classic chapters as 'My Pipe', 'My Tobacco Pouch', 'My Smoking Table', and pays homage to the Victorian smoker, while simultaneously disseminating advice on etiquette. The values of the middle class were perpetuated through such fables, as is demonstrated by Barrie's account of his discovery of the ideal pipe tobacco in a chapter of *My Lady Nicotine* entitled 'The Arcadia Mix'. This blend, whimsically named to evoke the rustic fantasies of the Elizabethan age, was claimed by Barrie to produce the perfect smoke, but Barrie kept its true identity a secret. To divulge where the blend might be purchased 'would be as rash as proposing a man with whom I am unacquainted for my club. You may not be worthy to smoke the Arcadia mix.' Readers were warned that the Arcadia mix, like a shaman's potion, could be dangerous in the wrong hands, for: 'This mixture has an extraordinary effect on character, and probably you want to remain as you are.'

The new middle class developed courtship rituals designed to allow its males time to earn enough money to maintain a family in appropriate style. This necessitated a period of bachelorhood before wedlock. Pipes were intimately associated with the portion of a man's identity formed during the bachelor istar. They were an essential tool of the male bonding that characterized the Victorian age, so much so that men were often forced by their wives to give up smoking upon marriage. This painful experience is narrated by J. M. Barrie in the first chapter of *My Lady Nicotine* entitled 'Smoking and Matrimony Compared'. The book's final chapter, after many pipe-smoking adventures, 'when my wife is asleep', finds the author with his empty pipe in his mouth, for although he has given up smoking, he had not surrendered his pipe. Victorians seemed to consider tobacco and women mutually exclusive, and in contests between them women did not always come off the winners. Rudyard Kipling's famous verses on this conflict demonstrate the dilemma faced by many bachelors:

> For Maggie has written me a letter to give me my choice between
> The wee little whimpering love and the great god Nick o' Teen
>
> And I have been servant of Love for barely a twelve-month clear,

But I have been Priest of Partagas a matter of seven
  year;
. . .
Open the old cigar-box – let me consider anew –
Old friends, and who is Maggie that I should abandon
  you?

A million surplus Maggies are willing to bear the yoke;
And a woman is only a woman, but a good cigar is a
  smoke.

The anguish of mutually exclusive choices appears to have
been peculiarly middle class. The poor smoked in front of
their women, indeed, poor women smoked. Members of the
upper class took a rather more cavalier approach to smoking
– and women – as the evil aristocrat Franklin Blake demonstrates in *The Moonstone* by Wilkie Collins:

Is it conceivable that a man can have smoked as long
as I have without discovering that there is a complete
system for the treatment of women at the bottom of
his cigar case? Follow me carefully, and I will prove it
in two words. You choose a cigar, you try it and it
disappoints you. What do you do upon that? You throw
it away and try another. Now observe the application!
You choose a woman, you try her and she breaks your
heart. Fool! Take a lesson from your cigar case. Throw
her away and try another!

The fictional identification of cigars with cads also appeared
in Thomas Hardy's *Tess of the D'Urbervilles* where wicked
Alec D'Urberville smokes a cigar whilst he watches the
delectable Tess eating strawberries, and contemplates her
seduction.

Sometimes, in the chaste if frustrated mind of the middle-
class bachelor, trying to make his way in the world before
settling down, the pipe was confused with sex itself. Witness
Mr Barrie's response when a female visitor to his bachelor

rooms cannot resist touching his pipe: 'Off tumbles the bowl. "Oh," she exclaims, "see what I have done! I am so sorry!" I pull myself together. "Madam," I reply calmly and bowing low, "what else was to be expected? You came near my pipe and it lost its head!" She blushes, but cannot help be pleased.'

Many men continued to smoke after marriage. The habit of dismissing women after dinner so that tobacco could be enjoyed in their absence became common in the nineteenth century, as did smoking rooms in private houses. The smoking room, whether situated in a country residence or the House of Commons, served as a sanctuary to which men retired to reminisce over tales of adventures overseas or take decisions of importance that required the absence of women and the consequent distractions of condescending to the weaker sex. The smoking room was idealized in the fiction of the period as 'that chamber of liberty, that sanctuary of the persecuted, that temple of refuge, thrice blessed in all its forms throughout the land, that consecrated Mecca of every true believer in the divinity of the meerschaum, and the paradise of the narghile'. Whilst the Victorians were aware of the importance of smoking in Amerindian cultures, thanks to the popularity of published works on the wild west, they would have resisted the suggestion that their gatherings in these smoke-filled rooms in any way resembled the tribal circle of braves, passing the sacred calumet around a fire, or indeed that their nascent pipe culture had any other foundation than British common sense.

In addition to using smoking as a tool of discrimination by class and sex, the Victorians also initiated age barriers for the practice. Smoking was a man's sport – something a child might aspire to, but should not enjoy until of age. This prejudice resulted partly from the 1870 Education Act which ensured that every British child was within reach of a school where they might learn to read, write and count. This and subsequent legislation altered the legal status of children. They became state charges as opposed to cheap labour, causing corresponding social changes in their treatment. Like women, they were deemed to require protection and moral guidance.

Boys who smoked were warned that they would grow up stunted, infertile or with 'smoker's heart', a debilitating condition that effectively excluded its sufferers from the glories awaiting them in the far-flung outposts of Empire. But boys will be boys and became aspirational smokers. They seem to have been encouraged in this course of action by many Victorian authors who included scenes of clandestine smoking in their works, and the youth of that age took to heart William Blake's maxim: 'Stolen fruits eaten in secret are sweeter.' The urge to smoke, a reflection, perhaps, of every child's fascination with fire, drove Victorians as young as three in quest of their first smoking experience:

> When I could elude the vigilance of a tiresome old nurse, I would sometimes enjoy the pleasure of pretending to smoke by sucking one of my father's empty pipes and expelling from my pursed lips imaginary fumes. That experiment must have been made first when . . . I was three years old, but the seductive aroma of those empty pipes is still a pleasant memory and one of the emotions I associate most vividly with childhood's yearning for the wonderful *au dela* of being grown up is the thought that one day I should be a real smoker.

Kim, the star of Kipling's eponymous novel, demonstrated that even children reared by animals shared the same desire to smoke, and was disciplined for smoking after leaving his jungle nursery for the white man's school.

To assist Britain's youth through its anguished smoke-free years, chocolate cigars and later chocolate cigarettes were introduced. These items of confectionery were manufactured to bear the closest resemblance possible to the real thing. Chocolate cigars were supplied with paper bands and tipped with edible fondant ash. In this manner, children were able to enact adult roles with placebos. They could brandish chocolate cigars over their ranks of tin soldiers or while playing with their toy guns in imitation of the military heroes of the age, whose histories were read aloud to them as inspiration. Curiously, the dangers posed to health by smoking were not presumed to carry beyond adolescence. It was believed that a grown man, like seasoned timber, could better withstand the shocks tobacco occasioned than a slender youth, a mere sapling, that might perish in its first drought or frost.

Opposition to adult smoking was limited, and such as it was, an offshoot of the temperance movement which considered poor people incapable of managing their pleasures, and therefore better denied them. In 1858, the Reverend Thomas Reynolds, an archetypal temperance figure, launched an 'anti-tobacco journal'. His motives were openly religious, and his prose is reminiscent of King James I's *Counterblaste*. Tobacco insulted God he argued: 'Not only is the use of tobacco an infringement to the laws of nature but it defeats the designs of our God, our Maker: and that which defeats his designs for purposes of mere sensuous gratification, must obviously be an affront to his Divine and excellent Majesty.' Reynolds was considered an idiot in his time and ridiculed accordingly. The sum he raised in subscriptions for his journal, £267, like its readership, was trivial. Copes, in a gesture of sportsmanship, offered Reynolds £1000 to stop publishing. His likeness appears in the figure earlier in the chapter, waving an umbrella of cant over the happy smokers 'livid with impotent spite and envy, utterly unheeded by the ardent votaries in ecstatic pursuit of our most gracious and glorious DIVA NICOTINA'.

A more serious dispute over the merits, and the possible disadvantages of smoking, was pursued in 1857 in the pages of *The Lancet*, a periodical founded to cater for the newish science of diagnostic medicine. The debate commenced when the eminent Samual Solly's 'Clinical Lectures on Paralysis' appeared in *The Lancet*'s pages, in which it was suggested that smoking may have been a contributing factor to the surge in cases of 'general paralysis' then being reported. Unfortunately, the medical issues were clouded by moral ones and the debate withered into a series of counter-accusations of godlessness. It seems Victorians did not believe tobacco was bad for them, except, like everything else, in excess.

They were, however, worried about their bowels, whose regular movements were believed to be the key to good health. Failure to evacuate promptly could lead to a disease named 'auto-intoxication', whose victims suffered the indignity of being poisoned by the contents of their large intestine. There was a giant Victorian market for laxatives to ward off this evil. The problem was simultaneously attacked from the other end with enemas, especially of tobacco. Although the clyster of the tobacco shaman had taken longer to reach British shores than the cigar, it arrived nonetheless, and the British began to use tobacco anally.

Victorian fondness for tobacco, whatever the manner of its use, was depicted in the literature of the age with the same wave of enthusiasm for realism that had inundated Holland's painters in the seventeenth century. No longer content with the literary equivalent of portraits of idealized princes, British writers aimed to describe life in every detail, unless sordid, in which case it was not described at all. This curious mixture of realism and prudishness was well received by the newly literate public. Charles Dickens, its most adept representative, filled his novels with tabagophiles, each with a habit appropriate to their class. Fictional characters were considered incomplete or unconvincing if they did not use tobacco.

Fictional tobacco use was stratified as much as in real life. Not only did classes' tobacco habits identify them like name badges, but their perceived reasons for smoking were different. The poor were shown to smoke to palliate their harsh and powerless existence. The middle class smoked reliably and unobtrusively, the minute differences of rank between them, like the cars of twentieth-century travelling salesmen, being expressed in their smoking apparatus and distinguishable only by themselves. The rich smoked effort-lessly, carelessly or frivolously, depending on how often they had travelled abroad, and whether their money was new or old.

Foreign women were included in this convention. Their usual literary role was that of temptresses, and a sure sign of a temptress was a tobacco habit. Men were presumed to lower their guard while smoking and female invasion of this male preserve indicated a familiarity that could only lead to trouble. In order to avoid moral conflicts amongst readers, foreign fictional female smokers usually got no further than tempting, expiring conveniently before the hero might succumb to the intoxicating combination of woman and weed. In Ouida's *Under Two Flags*, a best-seller of the age (whose story forms the basis of the film *Beau Geste*), the heroine of desert battles is named Cigarette. An early fictional example of nominative determinism, Cigarette smokes cigarette after cigarette. Cigarette was the closest the Victorians got to 'girl power' and this sexy little creature of the sands probably did more damage to the hearts of Britain's youth than her combustible namesake.

In addition to indicating when and by whom smoking might be appropriate, and what to expect if a character's tobacco habit was at odds with their sex, age or occupation, Victorian writers also formulated romantic and universal answers to the question of 'Why smoke?' Charles Kingsley, author of *The Water Babies* and *Westward Ho!*, following in the footsteps of other children's writers who celebrated the weed in print, presented tobacco's virtues thus: 'A lone man's companion, a bachelor's friend, a hungry man's food, a sad man's cordial, a wakeful man's sleep, and a chilly man's fire.'

Tragically, the majority of the new middle class lived lives of such stultifying dullness that the evocations of Elizabethan England and the Walter Mitty style fantasies served up by the literature of the age were necessary palliatives. To cater to the delusions of a largely inactive readership, Victorian writers evolved a peculiar hero, born in the ethos of amateurism, whereby innate ability always triumphed over plodding effort. This ideal was listless in everyday society, but could summon up fierce energy when circumstances demanded. Its perfect example was Sherlock Holmes, Sir Arthur Conan Doyle's Baker Street sleuth. Holmes is the quintessential Victorian adventurer, set loose in the urban jungle instead of the broad expanse of empire. 'My life is spent in one long effort to escape from the commonplaces of existence,' Holmes complains in the *Red Headed League*, a sentiment that would have rung true in many readers' breasts. Tobacco provides clues in many of Holmes's adventures, as he has studied smokers at length, even publishing a 'little monograph on the ashes of 140 different varieties of pipe, cigar and cigarette tobacco'.

Holmes, in keeping with his aura of mystery, smoked a wide variety of pipes and tobacco during the course of his adventures, and occasionally brandished a cigarette. His reasons for smoking were clear – it was a vital accompaniment to thought:

> 'As a rule,' said Holmes, 'the more bizarre a thing is the less mysterious it appears to be. It is your common, featureless crimes which are really puzzling, just as a commonplace face is the most difficult to identify. But I must be prompt over this matter.'
>
> 'What are you going to do, then?' I asked.
>
> 'To smoke,' he answered. 'It is quite a three pipe problem.'

The Victorians took their own peculiar notions of tobacco use overseas. They were moderns wandering into the medieval, or even biblical ages when they went abroad. At no other time in history was the discrepancy in technology between cultures so evident. The Victorians not only claimed to be God's chosen ones, they knew it, and it was a matter few other nations would dare dispute. The growing empire held opportunities for every class of Victorian society. Criminals were transported to it, the poor emigrated there, the middle classes were employed in administration or civilizing, and the upper echelon accepted appointments of a regal nature.

Following the example of the seventeenth-century Dutch, tobacco growing was encouraged throughout the British Empire, and the weed was cultivated in nearly every territory under the protection of the British flag. India, the jewel in the crown of Britain's overseas possessions, was the second largest producer of tobacco in the world by the third quarter of the nineteenth century, although almost all the weed it grew was for domestic use. From the Himalayas surrounding the newly named and measured Mount Everest to the subcontinent's southern tip, tobacco was consumed and celebrated in a variety of manners as diverse as the Hindu pantheon. Indians, however, unlike their colonial rulers, did not consider tobacco and women to be mutually exclusive. According, for example, to a Gadaba myth: 'there is no difference between tobacco and a wife; we love them both equally'.

The scale of Indian production was immense. For instance, in 1884, it harvested 340 million pounds of tobacco – four-fifths of the equivalent in the United States, then the world's largest producer. Most tobacco grown in India was dark and either chewed or, more commonly, consumed in the form of small cigars named 'bidis'. The water pipe, or hookah, was firmly established in the Mogul states, and had been adopted in a manner reminiscent of the Elizabethan 'reeking gallants' by members of the British administration, who employed special servants, termed 'hookah-burdars' to carry and tend the instrument of their pleasure. As Major General Keatings recalled in his memoirs: 'In former days it was a dire offence to step over another person's hooka-carpet or hooka-snake. Men who did so intentionally were called out.' Indian passion for tobacco possibly exceeded that of their white masters – a popular saying in Bihar – 'show me the man who can live without either chewing or smoking tobacco' – is illustrative of the absorption of tobacco into India's myriad cultures.

In addition to offering members of the British middle class

employment in trade or administration, the Empire held the temptation of exploration. Opportunities to be the first white man, or the first man at all, in one of the many empty spaces in the atlases of the day abounded. After the successful exploration of America's centre, Victorians burned to know what went on in the middle of other continents such as Africa and Australia. Was anybody there? Did they smoke? Even the terrain in these places was a mystery – were there mountains, forests, inland seas perhaps? Australia's interior held few of these, although its explorers gamely died of thirst while searching.

The Victorians had more luck with the 'Dark Continent'. Figures such as David Livingstone, Henry Stanley and Richard Burton prayed, fought or philandered their way into the heart of Africa, solving geographical puzzles along the way. All three noted the ubiquity of tobacco and found time to marvel at African pipe design, which, in terms of exuberance, was far in advance of the meerschaum or briar. While the explorers sometimes took time to describe the ceremony and importance attached to smoking among 'savage' tribes with amused condescension, they themselves were not loath to use pipes as decision-making tools. Before embarking on his perilous journey from the source of the Congo to the Atlantic Ocean, Henry Stanley consulted with his last living white companion over the prudence of the planned expedition. In his own words:

It was my after dinner hour, the time for pipes and coffee, which Frank was always invited to share.

When he came in, the coffee pot was boiling, and little Mabruki was in waiting to pour out. The tobacco pouch, filled with the choicest production of Africa, that of Masansi near Uvira, was ready . . .

'Now Frank, my son,' I said, 'sit down. I am about to have a long and serious chat with you. Life and death – yours as well as mine, and those of all the expedition – hang on the decision I make tonight.'

Putting Africa on the map

The figure above shows these two heroes at work, deciding the fate of their followers. Curiously the outcome of the second decision-making procedure they employed – coin tossing – predicted accurately the direction in which their voyage would end, resulting in glory for one, and death for the other.

In general, however, both exploration and Empire were serious businesses and hence the high mindedness advocated by the Victorians. Moral accountability was a governing motive of the era, which 'induced a highly civilised people to put pleasure in the background and what it conceived to be duty in the foreground'. This often resulted in gloomy self-denial as much as altruism or philanthropy. There was a passion for self and universal improvement as opposed to self-indulgence. It is surprising, therefore, that smoking did not attract more criticism in the latter half of the nineteenth century. It was undeniably dirty and smelly, and presented its aficionados with an ideal opportunity to practise wholesome self-restraint. The Victorians appear to have avoided this dilemma by resorting to the grounds of 'patriotic duty': 'The Chancellor of the Exchequer – the entire working of the machinery of government in this great country – the existence and efficiency of our army and fleet – largely depend upon the financial results of the consumption of tobacco by our truly patriotic smokers.'

However, not every Victorian allowed their life to be governed by such demanding precepts. Their actions were sometimes prompted by convenience, not duty, and the increased availability of tobacco brought many new smokers into the fold. A plentiful supply was matched by the number of places and occasions on which it was practical or permissible to smoke. Thanks to the invention of the friction match, smoking had become a more mobile habit. Stockton-on-Tees had the honour of its birthplace, a chemist named John Walker the laurels of its discovery in 1827. Walker was imitated by a London chemist named Samuel Jones who marketed his wares under the brand name 'Lucifer'. Although invented to assist smokers, the range of applications for a portable device to make fire clearly reached beyond lighting pipes and cigars. The British government, with its customary eagerness to profit from pleasure, and perhaps overconfident of its subjects' sense of duty, proposed a match tax in 1871. This time instead of claiming that matches were not strictly necessary to human existence, the tax was justified on the basis that matches were too cheap and therefore 'wasted in the most reckless and dangerous way'. The chancellor of the exchequer, Robert Lowe, proposed that the match duty stamp should bear the motto 'Ex Luce Lucullum' ('out of light, a little money'). The tax was abandoned after public outcry. England's principal match factory was within rioting distance of the House of Commons and its workers did not appreciate threats to their employment, even when decorated with classical witticisms. Further, the ragged match seller had become a fixture of street scenes and society rallied to support these decorative creatures' livelihood.

11

Automatic for the People

*The effeminacy of cigarette smokers, and
public reaction against them – the power
of advertising – cigarette production is
mechanized in America and Britain – the
practice of inhaling tobacco smoke revived
– monopolies and anti-smoking leagues*

Although convenience in the form of the friction match influenced when and where people might smoke, the principle had yet to be applied to what people smoked. Cigarettes were clearly more convenient than any other smoking device, yet for several decades after their introduction to Britain, they had an unsavoury reputation. They were tainted by their association with France, which no longer was an influence over British fashion. Memories of the guillotine were still fresh among Britain's elite, France had performed dismally in the Franco-Prussian war of 1870–1 and, as a consequence, things French (excepting its wines) were considered suspect. Cigarette smoking was frowned upon as a 'miserable apology for a manly pleasure', a habit that might only appeal to the 'effeminate races of the continent'. The few social or public figures who smoked cigarettes did so to shock. Oscar Wilde, flamboyant Irish homosexual playwright, horrified society by chain-smoking these effeminate devices, which he celebrated in his novel *The Picture of Dorian Gray*: 'A cigarette is the perfect type of a perfect pleasure. It is exquisite, and it leaves one unsatisfied. What more can one want?'

Pipe and cigar smokers closed ranks on the cigarette. These undersized, paper-wrapped tobacco sticks threatened the integrity of their own habit. The line of attack they selected, believing it to be most favourable to their cause, was health. Cigarettes, they claimed, quite unlike their own more robust ways of smoking, could only be appropriate to the anaemic office workers who seemed to be springing up in every British town, and whose feeble metabolisms were not up to a proper smoke. This theory was reinforced by cigarette smokers' obvious lack of manners, deductible from their insistence on trivializing the noble occupation of smoking. And no manners equalled bad breeding, therefore proving the original supposition that cigarette smokers were naturally inferior specimens and best shunned.

Similar opinions were voiced in the United States of America where the spreading cigarette habit filled a nation of tobacco ruminants with dismay. 'The decadence of Spain began when the Spaniards adopted cigarettes and if this pernicious practice obtains among adult Americans the ruin of the Republic is close at hand,' thundered the *New York Times* in 1883. Despite such vitriol, cigarettes were clearly tempting to American consumers. National cigarette consumption had grown from 42 million in 1875 to 500 million in 1880. American tobacco manufacturers took notice of this phenomenon. Why were so many people smoking cigarettes? Cigarette making had been a sideline for most tobacco manufacturers. Some would have nothing to do with it, believing that diversification into cigarettes might offend their existing customers. The manufacturers decided, after profound contemplation, that the answer to cigarettes' success was advertising. Whenever cigarettes had been advertised, their sales leapt.

Advertising required brands – a name or symbol representing a product to place in the public eye. Branding *per se* was then at a low stage of development in the United States. When people made purchases, they would ask for a measure or 'a dollar's worth' of goods, rather than a named product. Cigarettes, however, had excellent potential for branding as the result of a packaging innovation introduced in the 1870s by the leading tobacco manufacturer, Allen and Ginter. Hitherto, cigarettes had been sold loose or tied in bundles. Allen and Ginter began to sell their cigarettes in paper wrappers with a rectangular cardboard insert to stiffen the pack. The wrapper enabled Allen and Ginter to place their name under the customer's nose every time he reached for a cigarette. In addition, in order to provide the customer with the impression he had received something extra for free, the cardboard stiffeners were decorated with pictures of women, sporting stars and national monuments and issued in collectable series. Children clamoured for these cigarette cards and pressured their fathers into repeat purchases.

Emboldened by their success in regional markets, American cigarette manufacturers decided to harness the power of advertising to sell their wares throughout the United States. The first American cigarette manufacturer to have a stab at a national brand was W. T. Blackwell, which launched the first ever pan-American advertising campaign for its Bull Durham cigarettes in 1880. Adverts were placed in local, state and national papers and sales leapt. The proof that advertising could create demand made a strong impression on James 'Buck' Duke, the young head of a small Virginian tobacco company, who foresaw correctly that if demand was certain, the trick was to find a way to supply it.

Hitherto, smoking had escaped the Industrial Revolution. Most smoking devices were made by hand. Although mass consumed, they were not mass produced. Cigarettes were no exception. Tobacco factories were staffed with teams of Carmens, the most skilled of whom could turn out no more than five cigarettes per minute. In order to supply the demand he believed advertising could create Duke would need a workforce of millions. The principal cost of making cigarettes was labour. In the 1870s, the wages of rollers represented nine-tenths of the production cost of cigarettes. In order to advance, cigarette making had to be mechanized.

Many other goods, including chewing gum, razor-blades, soups and soap had already been subjected to this process, which brought the collateral benefits of consistency and cost savings. A machine could do more work more quickly and more cheaply than any team of humans.

The problem had been recognized by Allen and Ginter who offered a prize of $75,000 to anyone who could invent a cigarette-making machine. The challenge was taken up across the United States and in 1880 the 21-year-old son of a plantation owner named James Albert Bonsack obtained a patent for just such a device, which could make up to 212 cigarettes per minute. Allen and Ginter installed a Bonsack machine on a trial basis which, while it seldom managed to function at its optimum rating, spewed out an average of 70,000 cigarettes per day. At this point Allen and Ginter got cold feet. They rejected the machine after its trial period. Whether this was because they did not wish to pay the prize money, or simply were frightened of machines is not known. Buck Duke saw his opportunity and had two Bonsacks installed in his tobacco factory. Their inventor was determined that this time the machines would fulfil their potential and on 30 April 1884 one machine achieved its maximum rating for a ten-hour shift, producing 120,000 cigarettes. Duke immediately tied the Bonsacks to a favourable licensing agreement, which guaranteed his costs would be lower than any other manufacturer who used the machines, and therefore the lowest in the industry. Duke exploited this competitive advantage to the limit.

The answer to the question 'Why smoke?' for many Americans was simple: Buck Duke. He made cigarettes affordable and available and employed every marketing strategy in existence to ensure his products were in front of citizens throughout the nation. He pioneered tobacco sponsorship of sporting events with a roller-skating team called the 'Cross Cuts' who played in front of crowds as large as 12,000. He took up the idea of cigarette cards, although the sets issued by his brands, in his own words, featured either photographs or lithographs of buxom young ladies in what must have seemed very daring, if not shocking costumes. Usually these sets were labeled simply 'actresses' or bore descriptive phrases such as 'stars of the Stage', 'American Stars' or 'gems of Beauty'. Since there was surely little personal identification by the purchaser with the stars, who were usually unnamed, and since actresses were then accorded a low place in the social scale of polite America, it seems clear that such cards were designed for prurient attention.

Duke spent the then astonishing proportion of 20 per cent of his profits on advertising, which his technical advantage over other manufacturers enabled him to afford. Within five years of installing his first two Bonsack machines Duke was selling 2 million cigarettes a day, more every week than the French smoked in a year, and was making profits other manufacturers could only dream of. Duke's mission to bring his cigarettes to the people at a price they liked was assisted by a US federal tax cut on tobacco products, and by the fact that America had more people: its population doubled in the last two decades of the nineteenth century, the result of immigration and breeding. In the 1880s and 1890s there really was one born every minute.

A similar course of events had occurred in Great Britain, where the firm of W. D. & H. O. Wills began to operate Bonsack cigarette machines in 1883, after two of the firm's partners had been intrigued by the device at the Paris Exhibition of the same year. In 1884 they launched three machine-made brands of cigarettes: Three Castles, Gold Flake and Louisville, which within four years were selling over 11 million cigarettes annually. These initial brands' success took Wills by surprise. Hitherto, their most exciting market had been blended pipe tobacco. Why did so many Britons want to smoke cigarettes? Wills decided to continue the experiment and launched two more new brands in 1891 – Wild Woodbines and Cinderella, which sold 53 million and 32 million respectively in their first year of production.

As was the case in the United States, Wills passed on its cost savings from mechanization to the consumer, who also benefited from one of the few cuts in tobacco taxation in the history of the United Kingdom. This rare event took place in 1887, when the chancellor lowered tobacco duty by the princely sum of 4 pence per pound.

The reasons behind the exponential growth in cigarette sales on both sides of the Atlantic merit examination. Whilst the accepted cause in the United States was advertising, this did not apply in the United Kingdom. For instance, whilst Buck Duke had spent the then phenomenal sum of $800,000 on advertising in 1889 alone, the corresponding figure for Wills was closer to £10,000. Although the difference in expenditure is explained in part by the size of the respective markets and also by cultural anomalies – in Great Britain there was still a class-based resistance to purchasing goods on no other recommendation than a poster – it demonstrates that growth in cigarette sales was not merely a consequence of advertising. More and more people were smoking these cheap, machine-made devices, many of whom had never seen an advert.

Price certainly contributed to the cigarette's appeal in Britain. A pack of five Woodbines cost a penny – a sum that even children could afford. Cigarettes gave Britain's poor a cheap and accessible tobacco habit. The same criterion applied in the United States, where Duke was selling his brands at a nickel for a pack of ten. Cigarettes enabled more people to smoke more. In many cases they augmented existing tobacco habits. Annual tobacco consumption in the United States jumped from 2 to 3 pounds per capita in the last two decades of the nineteenth century.

Convenience was also a factor in the ascent of the cigarette. The populations of both the United States and Great Britain had become increasingly urban and accustomed to packaged goods, and the cigarette was the most convenient smoking device for the feverish pace of city life, where other forms of tobacco consumption were either too leisurely or unhygienic. Pipes took time to fill, longer to smoke and the aura of ease that had grown around their use worked against them. After the outbreak of tuberculosis in several American cities spitting came to be regarded as a health hazard, and metropolitan tobacco chewers switched to cigarettes. Cigarettes were ultimately portable. They gave pleasure quickly and consistently, could be purchased painlessly and required no further preparation or equipment other than a match. Cigarettes were also social. Whereas pipe smokers had developed a cult of individuality based on the principle of one man one pipe, and cigars were too bulky to carry round in numbers, or required storage in special conditions, a cigarette could be offered in an instant. It was as if each pack had the invisible command 'share me' printed on its surface. The final external factor in the cigarette's popularity was glamour. Cigarettes looked attractive and were smoked by elegant people. Their size was more evidently in proportion to the human frame than pipes and cigars and, like false finger nails, could decorate its appearance rather than encumber it. It seems tabagophiles no longer wished their smoking devices to be totemic. It was more important that they were easy and quick to use.

While advertising, cost, convenience and glamour all played a part in the growth of cigarette sales, the principal cause of their popularity was taste. Early hand-made cigarettes such as those produced by Philip Morris in his Bond Street salesroom had been compounded from Turkish or Egyptian tobacco, which were then considered the most suitable for the delicate palates of cigarette smokers. However, machine-made cigarettes used Virginian tobacco, in particular the charcoal cured 'yallacure' or Bright tobacco discovered by Stephen the slave on a farm in Caswell County, North Carolina back in 1839. Yallacure and other flue cured varieties of American tobacco were mild yet fragrant. They enabled a huge number of people who found pipes or cigars overwhelming to start smoking. The weakness of a cigarette in comparison to other then current forms of tobacco consumption became its strength.

There is also an essential difference between cigarette smoke, and that from a pipe or cigar. Cigarette smoke from charcoal cured tobacco is acidic, which hinders its absorption in the mouth and throat, but assists it in the lungs. In order for a cigarette to give the smoker a kick, its smoke has to be inhaled. Cigarette smokers revived inhalation in the developed world, where it had by and large died out. Unlike their counterparts in South America, Western pipe and cigar smokers had relied on the mucal train in their throats to convey the smoke's active ingredients into their system. Cigarette smoking therefore resulted in a fundamental change in smoking practice. People no longer smoked to stimulate their taste buds, but to irritate their lungs. This radical form of enjoyment – the repeated torture of vital organs – cannot be explained away by an explosive growth of masochism. Few people will accept pain as the price of pleasure. Smokers describe the sensation of inhaling as a tactile delight. The collision of smoke with lungs is an exquisite torture, and the practice of inhaling is as addictive as tobacco itself.

It seems, therefore, that the ever increasing number of cigarette smokers took up the habit because it was pleasurable, a pleasure aided by its cheapness and convenience. For perhaps the first time in the history of tobacco's association with man the weed was enjoyed simply for itself. The cigarette sheared tobacco use of its accumulated ritual and freed it from the encumbrances that had limited its prosecution. A cigarette could be enjoyed anywhere at any time.

In comparison to contemporary smokes, the first mass-produced cigarettes were, like the people of the age, shorter, broader and stronger. In the United States, whose population had sweet palates, the Burley and Bright tobaccos used in cigarettes were soaked in molasses and flavoured, like chewing tobacco, with a hint of liquorice. This treatment made the early cigarettes burn more slowly, and less evenly than those on sale today. By contrast, the British brands of the age had fewer additives. The British, it seemed, were content with the delicate flavour of tobacco solo. In terms of tar and nicotine content, cigarettes on either side of the Atlantic were formidably strong. This was balanced by the fact that people smoked fewer of them – five or ten a day. Some commercial brands were tipped with a band of cork to prevent the cigarette paper from sticking to the smoker's lip. This ingenious invention helped pipe and cigar smokers, who were accustomed to treating their smoking instruments more roughly, to enjoy a cigarette without the unlit end dissolving into a stew of tobacco, saliva and sodden paper. The cork tip's legacy survives in the decoration of the paper around a modern filter, whose tan colouring assists smokers in lighting the right end of their cigarette.

In the first five years of mechanized production there had seemed no limit to the amount of cigarettes America could smoke. The country's annual consumption in 1889 was 2.19 billion, a fivefold increase since the Bonsack machines had gone into production on Buck Duke's factory floor. And Duke had by far the largest share of this expanding market. It seems, however, that Duke was not content to be chief of the fastest growing and most profitable tobacco company in the USA. He wanted more; in fact, he wanted it all. His goal was the commercial idyll – a monopoly – which hitherto had been achieved by governments alone. Duke realized that for machine-made products big is beautiful and small manufacturers who were unable or unwilling to invest capital could be forced out of business. He started swallowing smaller competitors with a gusto that startled his principal rivals – Allen and Ginter, Kinney and, a name that was later to become famous in the tobacco world, R. J. Reynolds. Furthermore, Duke's aggressive pricing strategy was injuring all the major tobacco manufacturers. The president of Kinney declared he was 'most eager to get out of the advertising madhouse'. Major Lewis Ginter, patrician manager of Allen and Ginter, took the battle to Duke by offering to buy his company. But Duke, corporate rottweiler *par excellence*, was better placed to triumph in the ensuing dog-fight. He had the lowest production costs, the least principled sales team, and his morality was considered shocking, even in a business climate that valued cunning above integrity. In April 1889, Duke threatened to raise his advertising expenditure even further, which forced the hands of the other major producers. They met in New York in the same month and agreed to form the 'American Tobacco Company', a trust dedicated to becoming the sole supplier of tobacco products in the United States of America. The first president of this behemoth was the thirty-three-year-old Buck Duke.

Competition had also appeared in the British market. In 1883, Mr John Player of the Castle tobacco factory, Nottingham, had 'commenced the manufacture of cigarettes, and has adopted the names of Our Heroes, Our Charming Belles, the Castle Brand, etc.'. Players, however, thanks to Wills's foresight on insisting on exclusive British rights to the Bonsack machines, did not become mechanized until 1893. Wills was also challenged for market share by Lambert & Butler, the London based firm established as purveyors of cigars to the gentry, which mechanized its cigarette production in 1895, and, north of the border, by Stephen Mitchell & Sons, a Glasgow firm founded in 1723 and proud torchbearer of that now mighty city's tobacco heritage. Mitchell mechanized in 1899.

The British market was characterized by strong regional biases. Each manufacturer enjoyed a loyal local customer base and attempted to strike out at its rivals from this stronghold. In 1901, however, the internecine warfare was interrupted by the appearance of the American Tobacco Company. Its unexpected and unwelcome presence in what had been a domestic disagreement was the result of a totally unforeseen event in the United States – a fall in cigarette sales. The impossible had happened: America was saturated with cigarettes. The decline commenced in 1896 and continued until 1901. The American monster needed a continual supply of fresh consumers to justify its machines. If these were not to be found at home, then they had to be sought abroad. Although Great Britain looked small on the map, it was full of smokers, conveniently English speaking, and further benefited from a free tobacco market, unlike other European countries, where the trade was locked up in state monopolies.

Duke began his attack by pushing US brands in the same manner as he had advanced his own in the days before American Tobacco. He encountered stiffer resistance than he had anticipated. British consumers were not yet ready to surrender themselves to advertising. They also had the irritating tendency to associate price with quality, believing cheap things to be inherently inferior and therefore demeaning to their owner. Further, they possessed prejudices in taste and regional loyalties that appeared inexplicable to a man used to selling to the heterogeneous inhabitants of a country that had been virgin, or native, territory within living memory. In the parlance of the recently formulated game of baseball, Duke decided to 'step up to the plate' himself. He travelled liner, first class, arriving in England in September 1901.

Duke's first act on British soil was a demonstration of wealth, which equated to power back in the United States. He needed the British to know that he carried the biggest wallet in town. Before the month was out he had purchased Ogdens of Liverpool, a prominent regional manufacturer, whose most popular brand of cigarettes 'Tabs' were competitors of the market leader Woodbines. Interestingly, although Ogdens has long vanished into corporate heaven, its nineteenth-century brand name survives as a synonym for cigarette in the north of England, and amongst British soldiers.

The British closed ranks on Duke. Great Britain was still the greatest trading nation on earth, and neither its empire nor its markets had been won by principles alone. The thirteen largest British manufacturers decided to play Duke at his own game and combined to form the Imperial Tobacco Company, whose name alone should have set the Americans on their guard: if they wanted the world, it was well to remember who owned a third of it. Imperial Tobacco, headed by the 71-year-old Sir William Wills, counter-attacked Duke's domestic threat by offering to purchase any remaining independent US tobacco producer. It also bought up retailers, therefore shutting out Duke's brands. The war was short and culminated in a truce: American Tobacco and Imperial would have monopolies in their own markets and set up a third, jointly owned company named British American Tobacco, whose business was to exploit the rest of the world – a true multinational with a global remit.

While tobacco companies had been enjoying the fruits of mechanization and dividing the planet into territories, their product had drawn criticism as well as converts. Appropriately for a popular product, the cigarette was the first smoking device in history to attract popular opposition. Previously, tabagophobia had been the province of a saint, a king, tyrants, fanatics and oppressive governments. The cigarette, however, made numerous enemies among ordinary citizens, who attacked it on several fronts. Some relied on the arguments advanced by pipe and cigar smokers – that cigarettes were unmanly, fit only for the lips of Frenchmen and other lesser breeds. Others followed the religious path forged by the self-declared 'Devil's greatest enemy' King James I. Cigarettes, these pious tabagophobes reasoned (along familiar grounds), polluted man, God's best work and therefore were evidence of demonic activity. It is no coincidence that the beleaguered exponents of religion had suffered ever since their beliefs had been unmasked by Charles Darwin and felt obliged to seek proof of the Bible's continuing relevance.

The revival of the proposition that tobacco might be an instrument of Satan attracted the greatest attention in the United States, in particular in the 1890s when the approach of a new century was viewed with a degree of trepidation. Apocalypses were presumed to coincide with major calendar changes, and certain passages in the Book of Revelation appeared to have been written with 1900 in mind. The anti-cigarette movement was led by Miss Lucy Gaston, a spinster schoolteacher described as having 'a beardless resemblance to

Abraham Lincoln'. Miss Gaston possessed enthusiasm in
inverse proportion to her looks and held anti-cigarette rallies
in schools and temperance missions across the land. She also
published an anti-tobacco periodical called *The Boy* which
advised America's youth how to avoid the temptations of
cigarette smoking, and, if they succumbed, how to cure themselves of the consequences, which included 'cigarette face'.
Miss Gaston's enchanting blend of piety and pseudo-science
captivated America. The country had experienced an
alarming growth in the number of child smokers and the
epidemic threatened to become a plague. Cigarettes,
according to a temperance pamphlet of the time, were 'doing
more to-day to undermine the constitution of our young
men and boys than any other one evil'.

The temperance movement found an ally in J. H. Kellogg,
inventor of the cornflake, who was as obsessed with bowels
as the Victorian Britons had been. Kellogg, believed cigarettes were the most common cause of premature decrepi-
tude in America: 'Many a young man finds himself as old
at twenty or twenty-five years of age as he ought to be at
sixty or seventy. His constitution has been dissipated in
smoke at the end of a cigarette.' The American Congress
recognized an opportunity in this sentiment. Children, after
all, were the nation's future. In 1898, it raised tobacco duty
by 200 per cent to pay for the war against Spain. This was
the third time, including its war of independence, that the
United States had used tobacco to subsidize its conflicts.
Some states, including Iowa and Tennessee went further,
banning the sale of cigarettes altogether.

By the end of the nineteenth century, a similar movement
had emerged in Great Britain. Child smoking was on the rise
again, and true to the form predicted by pipe and cigar
smokers, juveniles preferred the unmanly cigarette. The
Reverend Reynolds had long since gone to meet his maker,
but other individuals and organizations emerged to carry the
tabagophobe torch. Instead of relying on religious objections
to cigarettes, these new opponents focused on the vulnera-
bility of youth. Britain's dangerous European neighbours,

jealous of its Empire because they wanted empires too, might at any moment subject the kingdom to a trial of strength. Its future soldiers, who would one day be called upon to do their patriotic duty, required protection in their tender years. They could not be allowed to risk the nation for cigarettes. A bout of pamphleteering commenced, which drew attention to the apocalypse that might result if the young men of Britain were not up to life's duty: 'If our beloved country is to escape the dire humiliation which recently surprised the French people, and if it is to continue to sustain a foremost rank among nations, its youth must forswear the indulgences into which too many recklessly plunge . . . lest one day some hostile nation should hurl confederated legions upon our sacred isle, and find Britain but a name, and her sons degenerate.'

Some of these fears seemed proven by the Boer War, where Britain's troops underperformed, not least of all because of their habit of advancing in highly visible scarlet ranks against mobile and camouflaged sharp-shooters. A newly returned veteran confirmed tabagophobes' suspicions that cigarette smoking may have been a contributory factor in his promisingly named training manual *Scouting for Boys*. Sir Robert Baden-Powell, himself a fervent smoker, explained to the nation's aspirant soldiers that smokers 'generally turned into rotters'. Sir Robert's manual further attempted to discredit what he considered to be the usual reasons for boys to want to try smoking: 'No boy ever began smoking because he liked it, but generally because either he feared being chaffed by the other boys as afraid to smoke, or because he thought that by smoking he would look like a great man – when all the time he only looks like a little ass.' The anti-youth smoking propaganda was supported by a Parliamentary committee on physical deterioration, which suspected that up to a third of men rejected for service in the Boer War were suffering from smoker's heart. The government took notice of this alarming statistic and in 1908 Parliament enacted the Childrens' Act, part III of which prohibited the sale of tobacco products to children of less than sixteen years of age. The prohibition was to be enforced by fines on retailers and conferred the power of confiscation on police officers and park keepers.

While the first decade of the new century had witnessed popular opposition to and public action against cigarette smoking on both sides of the Atlantic, these had been the result of religious hysteria and/or pseudo-scientific speculation, and had focused on the cigarette, not tobacco. Curiously, neither the British nor American anti-smoking movements paid any attention to advances in medicine and chemical analysis that had occurred contemporaneously, some of which suggested that tobacco posed genuine threats to the human metabolism. In 1889, two English scientists, Langley and Dickinson, published a landmark study on the effects of nicotine on the ganglia. They hypothesized that the nervous system was comprised of receptors and transmitters that responded to stimulation by specific chemicals, one of which was nicotine, which could therefore influence or interfere with the workings of the nervous system. In 1895, in Würzburg in Germany, Wilhelm von Röntgen discovered the X-ray, which enabled doctors to peer inside living bodies, and to learn the secrets of their bones and lungs. Knowledge of what smoking might do to its adherents, and how to look for it, was being acquired piece by piece. The evidence discovered thus far was alarming. In America, the pioneer botanical geneticist Luther Burbank observed that cigarettes were 'nothing more or less than a slow but sure form of lingering suicide' and their repeated use obviously detrimental to the bodies of their users: 'would you place sand in a watch?'

12

Of Cats and Camels

*Edwardian smoking – American Tobacco's
monopoly is found out and terminated –
the concept of addiction and the birth of
psychoanalysis – cigarettes enslave
America's youth – tobacco in the trenches
of World War I*

At the beginning of the twentieth century, tobacco's relationship with mankind had never been so healthy. Every known race of humans had a tobacco habit. The weed was present on all the continents bar Antarctica, where it was about to be carried by Captain Robert Falcon Scott. In the four centuries that had elapsed since Rodrigo de Jerez and Luis de Torres had sucked at the Cubans' giant cigar, tobacco use had become one of *Homo sapiens'* defining characteristics. Forests had been felled, or other crops cleared, to accommodate plantations of the weed. The relationship between human and plant had reached the stage of symbiosis. Symbolic, perhaps, of tobacco's triumph, was one of the first acts of Britain's new king, Edward VII, who appeared in front of a gathering of friends in a drawing room in Buckingham Palace after the death of his tabagophobe mother Queen Victoria in 1901, carrying a cigar, and announced 'Gentlemen, you may smoke.'

Tobacco consumption levels in the industrial world had leapt – the centuries-old average of approximately two pounds per head per annum had reached five pounds in the USA and was soon to be three in Great Britain. Although the cigarette had introduced many new smokers to the weed, traditional forms of consumption still prevailed. Cut tobacco for pipe smoking (or in the case of the United States, chewing) accounted for most tobacco sales in Europe and the USA, followed by cigars, with cigarettes and snuff making up the remainder.

The weed maintained its association with the arts. A new generation of British playwrights, including George Bernard Shaw, took to tobacco with the same enthusiasm as their Elizabethan counterparts. French artists such as Henri de Toulouse-Lautrec found beauty in the blue-grey whorls of cigarette smoke; Piet Mondrian composed his spare abstract paintings pipe within mouth, Vincent van Gogh his sunflowers similarly bedecked. The illustrator, Aubrey Beardsley, found a cigarette between his subject's fingers could enhance the proportion of the sitter and the surrealist movement explored the surprising symbiosis of man and plant. In some cases, the smoking implement itself came to the foreground: René Magritte isolated the pipe as one of the emblems of suburban man, suspending the cloud-maker against a plain background. The weed likewise retained its links with men of science. Albert Einstein, the greatest scientist since Darwin, mused over the mysteries of space and time with a pipe in his mouth. Einstein's reasons for smoking were clear: 'I believe,' he said, 'that pipe smoking contributes to a somewhat calm and objective judgement in all human affairs.'

All over the globe, people were reaching for pipes, cigars, snuff boxes and the newfangled cigarettes. China alone consumed 1.25 billion cigarettes in 1902. The Japanese had also taken to this most Zen of smoking devices, which consumed itself in fire for pleasure. Meanwhile, in the United States, the American Tobacco Company, the monster created by Buck Duke to corner the nation's cigarette market, was beginning to throw its weight around. Although it did not enjoy the punitive powers of European

Man's best friend

state-owned monopolies – it could not, for instance, imprison or kill its competitors – it had formidable financial muscle which it used to threaten any company or individual that stood in its way. For example, when a retailing chain attempted to sell competitors' brands they were advised by American Tobacco that unless they pushed its products alone 'we would consider it in our interest to hamper your enterprise in every way we could'. The retailer complied, knowing that American Tobacco could and would set up shops within sight of every one of its own outlets, selling cigars, cigarettes and chewing tobacco at half price until the retailer went out of business.

In addition to leaning on retailers, American Tobacco set to work on its suppliers, the tobacco growers in the Southern States. These countrymen, descendants of the citizens responsible for America's existence and independence, put up a stiffer resistance than its shopkeepers. They formed themselves into co-operatives to keep supply tight and prices reasonable. Tobacco farmers who broke ranks were ostracized or horse-whipped. Unhappily, they were too poor to hold out for long, although their plight attracted the nation's sympathy. American Tobacco, in a philanthropic gesture, settled the price it would pay for raw tobacco at nearly 8 cents per pound – enough to end bad publicity, to keep the farmers alive and growing tobacco, but no more.

American Tobacco also applied its market practices to politics. The age of mass production had proved to have unpleasant side effects. The food and pharmacological industries operated without regulation, and many of their products were rotten, contaminated or poisonous. To protect Americans the Federal Food and Drugs Act was passed by Congress in 1906, which prohibited the sale of adulterated foods and drugs, and decreed that an honest statement of contents should be made on labels. Originally, nicotine had been on the list of drugs to be subject to the Act. However, after American Tobacco lobbied congressmen with dollars and the argument of princes that tobacco was neither a food nor a drug it was removed from the list.

American Tobacco's party was spoiled by the twenty-sixth president of the United States, Theodore Roosevelt, whose motto was 'speak softly but carry a big stick'. Other companies were attempting monopolies in the oil and railroad industries and Roosevelt was determined to halt the trend. Anti-trust, i.e. anti-monopoly, proceedings were commenced against American Tobacco in 1907. They were vigorously defended. Buck Duke, now in his fifties, spoke candidly of his company's wish to gain market share, which was, after all, one of the basic aims of capitalism. He also pointed out that consumers had yet to suffer – tobacco products were cheaper and better than they had ever been. And as for choice, American Tobacco had nearly a hundred brands. What more could the public want? His opponents took a different view of American Tobacco's aspirations: 'All it wants is the earth with a barbed wire fence around it. The Tobacco Trust [American Tobacco] is a hog and wants all the swill.'

The decision went against Duke in May 1911. The US supreme court judged that American Tobacco had aimed for 'dominion and control of the tobacco trade, not by the exertion of the ordinary right to contract . . . but by methods devised in order to monopolize . . . [and] by driving competitors out of business'. The monster was dismembered three ways and Duke retired from the tobacco business. A substantial part of his fortune was dedicated, in the best of American traditions, to charity. Little Trinity College in Durham, North Carolina, was the Cinderella he selected to transform. The resulting princess was renamed Duke University.

After segmentation came competition. The three companies created from American Tobacco, and the few surviving independent competitors, looked carefully at the US tobacco market and set off in different directions. Some considered chewing tobacco to be the future, others that a pipe mania would soon sweep over America. R. J. Reynolds, whose Prince Albert pipe tobacco with a picture on the tin of Edward VII and the motto 'Now King' had been one of the few brands to compete with any success against American Tobacco, decided to bet itself on cigarettes, specifically, on a new brand. Before embarking on this project, in a rare display of corporate solicitude, R. J. Reynolds investigated the health claims that had been made against cigarettes. Despite the fact that cigarette sales had recovered from their *fin de siècle* slump, the little white sticks had continued to attract public criticism. For example, in 1909, the baseball star Honus Wagner had ordered the American Tobacco Company take his picture off their 'Sweet Caporal' cigarette cards, fearing they would lead children to start smoking. The shortage made the Honus Wagner cigarette card the most valuable of all time, and children hoarded their pocket money or pestered their fathers to buy Sweet Caporals in the hope of obtaining one. The card is now worth nearly $500,000. Only after three separate laboratories had reported back to RJR that cigarettes were risk free did the company proceed with its new project.

The most important issue RJR tackled was its new brand's name. Initially, in view of the success of 'Prince Albert', another representative of foreign royalty was considered. Germany's Kaiser Wilhelm was the obvious candidate, but was rejected on two counts. First, Edward VII who adorned the tin of Prince Albert had had the ill grace to die, leading competitors to suggest that the brand's slogan 'Now King' should be altered to 'Now Dead', and second, as RJR's founder observed, the problem with selecting a living figure was that 'you can never tell what the damned fool might do'. In hindsight, as the events of 1914–1918 were to prove, dropping the Kaiser was a lucky decision. RJR turned its attention from royalty to animals. Across the Atlantic in Britain, a brand named Black Cat was having considerable success Clearly cats have nothing to do with tobacco or smoking, but the absence of a link had not prejudiced the British public, who seemed to enjoy the picture of a feline on their cigarette packets.

Americans did not share the British obsession with cats, but it was thought that they might be susceptible to the charms of another animal. The recipe for RJR's new ciga-rette had already been formulated – a blend of Bright and western Burley with a touch of Turkish tobacco, whose pres-ence RJR wished to emphasize, as oriental cigarettes then carried an aura of sophistication. The association with the exotic East provided inspiration and the new cigarette was named Camel. Its pack bore a picture of the animal, in fact a dromedary, set against a background of desert sands and distant pyramids. Camels were a runaway success – the flavour was popular, and, moreover, the branding seemed to have cracked the American psyche. Smokers spent hours scrutinizing the packaging, in particular the picture of the camel, for other hidden images, including the figure of a naked woman.

In a manner which tobacco's surrealist admirers would have approved, we must move from Camels via eels to an exam-ination of the one quality of tobacco that had remained a mystery throughout the weed's history in the West. It had a strangely compelling effect upon its users – after the briefest of introductions, it was a friend for life. People who wished to renounce their tobacco habits found it more than a matter of willpower. Why did tobacco become irresistible? Why would smokers forgo food, alcohol and even sex for tobacco?

The quality had been noted by Oviedo, King James I, and numerous other commentators over the centuries. With the arrival of the cigarette, the problem became more acute. Cigarettes seemed to possess so powerful a hold over their users that they controlled them. Smokers appeared to order their day, and all their actions, around cigarettes. Smokers were even seen indulging on the deck of the *Titanic* as it slid under the waves. The suggestion that a man could be enslaved by burning leaves would have required a brave face prior to the cigarette. But once the tobacco had been wrapped and could no longer be seen to be nothing but the cured leaves and stems of a plant an aura of mystery developed around the hidden contents. Thomas Edison, the scientist who tamed and sold electricity, reflected a section of public opinion when he decided that cigarettes were uniquely dangerous among tobacco products, on account of the paper wrapper which had 'a violent action on the nerve centers, producing degeneration of the cells of the brain, which is quite rapid among boys'. Edison, a compulsive tobacco chewer ('tobacco acts as a good stimulant upon anyone engaged in laborious brain work'), was so convinced of the danger of cigarettes that he refused to employ cigarette smokers. There were many other testaments of the cigarette's uncanny powers of insinuation. They were like the worst kind of whore – easy for everyone, but once they had been tried, no one could ever forget them nor stop wanting them.

The irresistibility of cigarettes became a public problem in America, and therefore a potential source of profit. Sears, Roebuck & Co., a trans-American mail order company, that had been set up to sell the young nation's pioneers everything they might need, from nails, ploughs and Winchester rifles to lace and toys, advertised a 'Sure Cure for the Tobacco Habit' in its famous catalogue, under the slogan 'Tobacco to the Dogs'.

It was at about this time that the infatuation afflicting smokers acquired a name. Some other plant extracts, notably nicotine's fellow '-ines' such as morphine, had been observed to have a startling effect on the personalities of their habitual

users. The conduct of opium fiends was compared to that
of debtors by the late Victorians, and they became known
as 'addicts'. The word was a legal term, 'addictus', meaning
someone who had been sentenced to servitude to satisfy a
debt. It did not, however, imply that an addict was a victim.
Their enslavement was a consequence of their own impru-
dence, or excess, and while capable of evoking pity in char-
itable breasts, was usually considered indicative of innate
weakness of character.

The term 'addict' passed into common usage where it was
applied to other forms of obsessive or compulsive behaviour, including sex and smoking. It was perjorative at first
-- to be called an addict was to be insulted. An addict was
weak, an addict lacked willpower, an addict had forfeited
his dignity as a man. Smoking, however, was so common,
and enjoyed by so many eminent individuals whose integrity
in other matters was so evidently beyond question, that the
term 'addict' was only applied to smokers by the most fanatic
tabagophobes, and then only in the context of juvenile cigarette smoking.

Addicts, as a class, were generally left to their own devices.
Most drugs were legal and freely available at the beginning
of the twentieth century. The liberal ideals that had prepon-
derated in the Victorian era held adults to be rational creatures, capable of making choices, for good or evil, and only
when their behaviour threatened other people or their property should it be circumscribed. If consenting adults became
infatuated with narcotics to their own detriment, that was
their problem, and society did not feel itself obliged to limit
the sources of temptation, nor free victims from their servitude. Addiction was incurable, and probably a useful mechanism for removing the inadequate from the gene pool.

Help, however, was at hand for this kind of moral leprosy.
Western medicine was advancing in leaps and bounds, identifying the causes of and in some cases discovering protection against mankind's oldest and most familiar enemies,
such as smallpox and tetanus. Instead of blaming diseases
on miasmas, or humoral imbalance, micro-organisms had

been discovered to be the culprits, which could be combated with the weapons of sanitation and inoculation. Doctors had also turned their attention to diseases of the mind, which although they had not yet been attributed to germs, if their causative mechanisms could be discovered, cures might be effected. The principal problem with such ailments was detection. Obvious cases of lunacy or idiocy aside, mental illnesses manifested themselves in the character of the sufferer, and not his person, and as the surrealists had pointed out, who is to say who is mad? Brain surgery was still in its infancy, and, besides, no one was sure what took place where in that puzzling mass of jelly, which showed up in the new-fangled X-rays as no more than a shadow inside the skull. The new science of electrotherapy was claiming some successes – it had proved that powerful electric shocks had long-term effects on brain function, and had even cured – or tamed – some lunatics. Such limited advances aside, the causes of brain malfunction, including the debilitating ailments of hysteria, neuroses and addiction, which were then occurring in epidemic proportions, remained unexplored.

Salvation appeared in the form of a young Viennese medical student by the name of Sigmund Freud, whose adventures in the unseen regions of the mind commenced in 1876 when he was sent to the Italian seaside town of Trieste to discover whether eels had testicles, a matter which science had previously overlooked. The endless dissection of marine life this quest involved began to disturb Sigmund: 'even in my dreams, in my thoughts,' as he confided to a friend, 'nothing but the great problems connected with the words "ducts", "testicles" and "ovaries" . . .'. When Freud returned to Vienna he decided to devote his career to the workings of the mind, and the investigation of the afflictions that tormented it.

Freud had his first crack at the grey matter with cocaine, which he had found could cure opium addiction if prescribed in the right doses. The drug also seemed to help people to talk freely about themselves, which gave Freud another clue:

speech, not a scalpel, would unlock the secrets of the human mind. Tragically, after one of his experimental cocaine patients committed suicide, Freud was forced to renounce cocaine as a diagnostic tool. He passed on his discoveries to the dental trade, which still uses the drug as a local anaesthetic, and continued his personal use of the ancient Incan stimulant, as he revealed in a love letter to his wife-to-be: 'Woe to you, my princess, when I come. I will kiss you quite red and feed you till you are plump. And if you are forward you shall see who is the stronger, a gentle little girl who doesn't eat enough or a big wild man who has cocaine in his body.'

There had to be some other key to the mental processes that caused anti-social or self-destructive behaviour. Freud was searching for a single answer, and in 1895, via a process that has never been clearly explained, discovered the prime cause for nervous afflictions. Just as the Victorians had perceived the greatest threat to their health originated in their bowels, Freud decided that nervous disorders, including compulsive behaviour, had their origin internally. This time, however, there were no guilty organs, but a sexual force named the 'libido', which people neglected at their peril. The libido was the impulse that made smokers look for naked women in the camel depicted on a cigarette pack.

Such disorders were curable through a process Freud invented named 'psychoanalysis'. This involved lengthy one-on-one consultations between the analyst and patient, during which they would explore key episodes in the patient's past, searching for some form of sexual denial or abuse in childhood, which once identified and acknowledged would remove blockages to the free flow of the libido, and thus cure the patient. Ailments capable of treatment by psychoanalysis included obsessive or compulsive smoking. In *Three Essays on the Theory of Sexuality*, Freud argued that boys 'who retained a constitutional reinforcement of lip eroticism had, as adult men, a marked desire for smoking'. Lip eroticism was curable, provided the eroticist underwent psychoanalysis and identified its cause. Freud was at ease with his own lip eroticism, smoking up to twenty cigars a day. He described tobacco as his 'sword and buckler in the battle of life', and used it to channel his own libido, that monstrous force which, if not properly controlled (and suitably acknowledged), might turn even a Viennese professor into a nervous ruin.

However, insisting that everyone with behavioural problems had been sexually abused in their youth became a difficult argument to maintain, in particular because Freud's patients came from respectable families. He, therefore, refined his theory: the abuse need not be actual – it could be dreamed, in fact, it had to be dreamed, because every little boy's libido had as its goal the murder of his father and the seduction of his mother, vice versa, of course, for girls. This tendency became known as the Oedipus complex.

The philosophical repercussions of Freud's theories were immense. He shifted the blame from bad blood to a bad upbringing. Your parents, in a sense, were still to blame, but their work could be undone by other means than the gallows or newly invented electric chair, sponsored by Thomas Edison. Freud's work succeeded in paving the way for a new perception of addicts, smokers included. If addicts were curable, then, like any other invalids, they were to be pitied instead of scorned. This change in status is the foundation stone of the contemporary school of thought that considers smokers victims instead of consumers, and tobacco companies heartless merchants of death.

Freud's theories were slow to spread. The child abuse, real or imagined, that was central to his message, attracted the wrong sort of disciple – indeed, some were expelled from his inner circle for allowing their libidos off the leash when alone with their patients. Sex was also a handicap to the advance of Freud's theories in the USA. When he was invited to speak there in 1909 (where he resolutely ignored No Smoking signs) the Puritan strain was in the ascendant. Sex was as bad as cigarette smoking, against which opposition had continued to mount. Americans appear to have adopted the Victorian theory of the mutual exclusivity of smoking and marriage. They, too, were elevating their women to levels of unnatural – and unasked for – purity. Smoking was depicted in American writing as a dirty habit, or a weakness, neither of which characteristics a woman wanted in her man. Penitent smokers were reformed before marriage in the fiction of the age, a conversion achieved by a combination of persuasion and feminine charm.

Victorian strictures against adolescent smoking had also emerged in the United States. Thomas Edison was joined in his crusade against cigarettes by other American captains of industry, including Henry Ford, the car manufacturer, who composed a pamphlet entitled *The Case against the Little White Slaver*, addressed to 'My friend, the American Boy'. Ford's case was a charge sheet listing the offences smoking had committed against America's youth, supported by the testimonials of leading figures in sport, science and industry. Cigarettes were accused of ruining adolescent hearts, athletic performance, powers of concentration, and of leading smokers on to harder drugs: 'Morphine is the legitimate consequence of alcohol, and alcohol is the legitimate consequence of tobacco. Cigarettes, drink, opium, is the logical and regular series.' Tobacco's deadly potential was demonstrated in *The Case against the Little White Slaver* by the killing of a cat. Its executioner was the president of the Anti-Cigarette League of America, who injected his victim 'with a hypodermic syringe full of . . . tobacco juice. In a few minutes the cat began to quiver, then tremble, then it had cramps, and in less than twenty minutes it died in violent convulsions. The poison destroyed the nine lives a cat is popularly supposed to possess.' Amongst his authorities, Ford included the opinion of Hudson Maxim, who 'has won world renown as the inventor of high explosives for use in battleship guns and torpedoes and for various other purposes'. Maxim rated tobacco as more deadly than the weapons of destruction his business produced:

The wreath of cigarette smoke which curls about the head of the growing lad holds his brain in an iron grip which prevents it from growing and his mind from developing just as surely as the iron shoe does the foot of the Chinese girl.

In the terrible struggle for survival against the deadly cigarette smoke, development and growth are sacrificed by nature, which in the fight for very life itself must yield up every vital luxury such as healthy body growth and growth of brain and mind.

If all boys could be made to know that with every breath of cigarette smoke they inhale imbecility and exhale manhood, that they are tapping their arteries as surely and letting their life's blood out as truly as though their veins and arteries were severed, and that the ciga-rette is a maker of invalids, criminals and fools – not men – it ought to deter them some. The yellow finger stain is an emblem of deeper degradation and enslave-ment than the ball and chain.

All debate as to the merits of cigarettes was swamped by events. In the late summer of 1914, Austria declared war on Serbia, whose ally Russia mobilized, followed by France, against which pair Prussia commenced hostilities, attacking France through Belgium, thus precipitating Great Britain's involvement on the side of the Belgians, the Russians and the French, and launching World War I. Even in the earliest days of the conflict, when the moods of the combatants were optimistic, tobacco was acknowledged to be a special need of the fighting man. Cigarettes and cigars were distributed by the Red Cross to Austro-Hungarian soldiers leaving for the front, and provided, in the form of tobacco rations, to combatants of almost every participating nation. In 1914, the British infantryman's tobacco ration was 2 ounces per week, while the German's daily ration was two cigars and two cigarettes or 1 oz. pipe tobacco, or $9/10$ oz. plug tobacco, or $1/5$ oz. snuff, at the discretion of his commanding officer. Britain's sailors and marines did better than the foot soldiers:

the Royal Naval tobacco ration was 4 ounces per week. Officers in the trenches often supplemented their troops' rations by buying cigarettes out of their own pockets. Siegfried Sassoon, MC, in his *Memoirs of an Infantry Officer*, records such a purchase on the way up to the front line: 'A large YMCA canteen gladdened the rank and file, and I sent my servant there to buy a pack full of Woodbines for an emergency that would be a certainty . . . twelve dozen packets of Woodbines in a pale green cardboard box were all I could store up for the future consolation of B Company, but they were better than nothing.' The French too took their *bruyère* pipes and '*grisettes*' with them to the trenches. *La Baïonnette* (The Bayonet), a French combatants' newspaper, regularly published tobacco litanies: prayers to tobacco written in gratitude and praise.

The war quickly settled down to one of attrition, with opposing armies facing each other across trench parapets, occasionally launching assaults whose results were measured in casualties rather than territory gained. The importance of tobacco to the fighting man in this environment was recognized by their governments, by aid organizations, by their relatives at home, who usually included cigarettes in comfort parcels, by the media, who launched tobacco funds, and by the troops themselves. Cigarettes were among the very few branded products, weapons aside, on the front line. Some brands became part of trench slang. For instance, the elderly Russian five-funnelled cruiser moored off Gallipoli was known as the 'packet of Woodbines'. In the light of the ambivalent reputation smoking was gaining prior to the war, the universal approval of the weed marks a significant shift in attitude. Why was the habit that 'generally turned men into rotters' now considered indispensable?

Combatants smoked for physical, social and spiritual reasons. Tobacco's mild narcotic properties, and its ability to suppress hunger, were valuable qualities in the horrifying conditions that prevailed in the trenches. Before a potentially fatal assault the act of smoking a final cigarette granted the soldier a moment to calm his nerves and prepare for the fighting ahead. Not only tobacco, but the act of smoking was important. The comfort of being able to perform a normal sequence of actions, such as the soldier might have carried out at home, had a soothing effect. Even the presence of smoking implements could help steady a soldier's nerves. A pipe might serve as a link from the front line to the country for which they were fighting. The following excerpt from Sassoon's *Memoirs of an Infantry Officer* demonstrates this role as Sassoon crawls through no-man's-land to rescue a wounded comrade:

> I could hear the reloading click of rifle bolts on the lip of the crater above me as I crawled along with mud-clogged fingers . . . I knew that nothing in my previous experience of patrolling had ever been so grim as this, and I lay quite still for a bit, miserably wondering whether my number was up: then I remembered I was wearing my pre-war raincoat; I could feel the pipe and tobacco pouch in my pocket and somehow this made me feel less forlorn.

The sociable nature of smoking proved to be as important as the physical effects it produced. Cigarettes were shared between friends and enemies alike. They were the currency of compassion amongst fellow sufferers. The offer of a cigarette was a token of humanity – one of the few available to men in such wretched conditions. The shared cigarette provided the last example of civility before savagery, calm before storm, kindness before brutality. It is no accident that the tradition of offering a last cigarette prior to execution by firing squad developed as a means of humanizing one of the most bestial aspects of war, providing comfort for both victim and executioners alike.

Tobacco's spiritual importance to the men on the front line is harder to define. In so many ways, a cigarette was a representation of their own existence – a short-lived, expendable item, transformed by fire into spirit and ashes. It was a symbol of mutability, like the blood-red poppies that

sprouted every summer in no-man's-land. It was also a token
of security, however transitory. Cigarettes were smoked in
minutes of rest, in periods of calm, and so were associated
with such moments. The freedom to smoke a cigarette
implied that the combatant had survived – for now.

When America entered the war in 1917, and for the
first time in four centuries the flow of humanity across
the Atlantic was reversed, the importance of tobacco to
combat troops was taken for granted. As General Pershing
(1860–1948), supreme commander of the US troops
observed, 'You ask what we need to win this war, I answer
tobacco, as much as bullets. Tobacco is as indispensable
as the daily ration. We must have thousands of tons of it
without delay.'

Tobacco rations for the US troops were provided predominantly as cigarettes and converted many soldiers to this
method of consumption. The US War Department purchased
cigarettes from manufacturers in proportion to their pre-war

market share. The principal beneficiaries were RJR's Camels and American Tobacco's Lucky Strikes. As had been the case for Woodbines, their names entered trench slang. A popular slogan as the American forces arrived at the front was 'the camels are coming', and the green of the US troops' uniforms was commonly described as 'Lucky Strike' green. The cigarette manufacturers displayed appropriate patriotism in their domestic advertising, drawing attention to their involvement at the front. In 1918, when the War Department bought the entire output of Bull Durham tobacco, the brand adopted a new slogan to celebrate: 'When our boys light up, the Huns will light out.'

The American 'doughboys' threw themselves into combat with a vigour that astonished their war weary allies. Like the French *poilus*, the tommies, the Anzacs and the Canadians, they discovered the cigarette's use as comforter and friend. The following extract from the diary of Floyd Gibbons, one of the first US marines into combat on the Western Front, describing his arrival at a field hospital after being badly wounded, provides an illustration of the cigarette in its essential World War I role:

> On my left side I was completely bare from the shoulder to the waist with the exception of the strips of cloth about my arm and shoulder. My chest was splashed with red from the two body wounds. Such was my entrance. I must have looked somewhat gruesome, because I happened to catch an involuntary shudder as it passed over the face of one of my observers among the walking wounded and I heard him remark to the man next to him:
>
> 'My God, look what they're bringing in.'
>
> Hartzell placed me on a stretcher on the floor and went for water, which I sorely needed. I heard someone stop beside my stretcher and bend over me, while a kindly voice said:
>
> 'Would you like a cigarette, old man?'
>
> 'Yes,' I replied. He lighted one in his own lips and

placed it in my mouth. I wanted to know my bene-
factor. I asked him for his name and organization.

'I am not a soldier,' he said . . . 'My name is Slater
and I'm from the Y.M.C.A.'

That cigarette tasted mighty good. If you who read
this are one of those whose contributions to the
Y.M.C.A, made that distribution possible, I wish to
herewith express to you my gratefulness and the gratefulness of the other men who enjoyed your generosity
that night.

13

Stars

Working-class smokers – women learn to smoke – tobacco as a messenger of love between the sexes – It Girls and smoking for health – tobacco appears in movies and is awarded allegorical roles

Most of the men who survived the trenches were confirmed smokers. They found the habit had also flourished in the homelands they returned to. Civilians' nerves had also suffered in World War I, and cigarettes sales had leapt accordingly. The old prejudices attached to cigarettes – that they were risqué or effeminate – had vanished in the battlefields. Opposition in Great Britain to the device that could unman men was quickly forgotten. By the end of the war a majority of UK tobacco sales were in the form of cigarettes.

The illusions of an ordered society in which every citizen had their place were destroyed by the war and were no longer paraded in British cigarette advertising. Post-war promotion necessitated a different mythology, a fanfare for the common man. The heroes of World War I had been soldiers, not generals, the tommy in the trenches, not the old man on a horse. Instead of royalty or statesmen, post-war cigarette advertising chose ordinary men in ordinary occupations as its role models – steel workers, crowds at football matches – and instead of highlighting individuality or discrimination amongst their customers, tobacco manufacturers sought to make a virtue of their drudgery by celebrating their industry. It was the working classes' turn to grow itself a new identity, and cigarettes were part of it.

Cigarettes also began to gather aficionados from the half of humanity who had previously been dissuaded from their use. Prejudices against women smoking had been strong prior to 1914. Britain's Religious Tract Society had warned women that smoking would result in them growing moustaches as a result of 'constant movement of the lips'. Although poor women had always smoked, respectable women who imitated them were threatened with a loss of status: 'they are simply assimilating themselves to this old Sally and that ancient Betty down in the dales or mountain hamlets, or to the stalwart cohort of pit-brow women for whom sex has no aesthetic distinctions'. However, during the course of the war many British women had taken over work normally carried out by men – they had effectively run the economy while their husbands, sons and lovers fought. This accustomed them to financial independence. They had money to spend, and some of it went on cigarettes. Prior to the war, the largest field of employment open to single women was domestic service, where they were often expected to observe the convent rituals of poverty, chastity and obedience. However, at the end of 1918, when the munitions industry wound down, only 125,000 of the pre-war total of 400,000 returned to serving tea and making beds. It seems that freedom was as compelling as cigarettes, and once women had tasted it in the war years, they wanted more. They also had aspirations on other male preserves, not least of all the right to vote.

Once this right had been granted to them, more and more women began to smoke cigarettes. In 1921, their consumption reached recordable levels – 300 tons. British men, by contrast, consumed over 67,000 tons of tobacco products in the same year. From this cautious start, female tobacco consumption rose steadily over the 1920s, almost all of it in the form of cigarettes. It appears that women smoked for different reasons to men. Few imagined themselves one of a circle of seated braves, or as a bearded philosopher surrounded by leather-bound books, when they lit up. Even fewer had smoking rooms at their disposal. Certainly many adopted the habit as symbolic of the other male strongholds they had wished to invade. Not every woman had been able to chain herself to railings in Westminster, but they could demonstrate solidarity by smoking. And if nothing else, it was a good way of meeting and choosing men.

Over 10,000 dance halls opened in the aftermath of the Great War which Great Britain's single women attended without shame. At these palaces of pleasure, the cigarette assumed the role of messenger between the sexes. Cigarettes were offered by men to women as a token of courtship – in the absence of flowers in a metropolis, the weed was the next best thing. Tobacco thus fulfilled the promise it had demonstrated the previous century among the *cigarerras* and *caballeros* of Andalusia and the languid whores and chroniclers of decadence in Paris.

Women found that the act of smoking a cigarette could be choreographed to enhance their sexuality and to telegraph desire. Despite Freud's discoveries in the field of 'lip eroticism', men generally limited the actions of their lips and tongue when inhaling. Women, however, could exaggerate these movements without embarrassment. After accepting a dance hall cigarette, their lips would caress its tip before they drew the smoke deep inside their bodies. Not only their mouths were busy. Smoking is more than sucking and blowing. The entire diaphragm is involved – the chest can swell, the stomach contract – not only the face but the torso is at the smoker's disposal for expressing emotions.

Women did not wait to be offered cigarettes – they started buying them themselves, a trend which manufacturers viewed with a mixture of excitement and trepidation. Should they experiment with special feminine brands to tempt the weaker sex, or would they respond, like their men-folk, to pictures of bare-chested labourers in primary industries? The manufacturers were faced with a conundrum. Female smokers represented a huge, untapped market, but advertising directed at them specifically might alienate their male customers, who would resent the idea of smoking a ladies' brand. This problem had been suffered by Carreras's Black Cat brand, which developed a reputation as a ladies' cigarette, though whether women had adopted them for their mildness or as strays was unclear. A fortunate social phenomenon solved the manufacturers' dilemma. Working women were not the only members of their sex in quest of freedom. The rich, young and frivolous also wanted a share. The daughters of some of Britain's most venerable families began to frequent public places after dark without chaperones. Their exuberant behaviour was chronicled by the newspapers who christened them the 'It Girls'. It Girls chain smoked. They favoured cigarette holders, which set off their fashionably slender figures. They also perfected the trick of timing exhalations to punctuate conversation with smoke instead of words, thus completing the transformation from smoke symbolizing thought to acting as its substitute.

Since It Girls exercised a powerful fascination over both sexes, their images could be used to promote cigarette smoking without offending either. Posters of smoking It Girls mark the appearance of fantasy in tobacco advertising. The common man and woman had their dreams and the cigarette manufacturers began to cater to them. The It Girl phenomenon also made its debut in the literature of the period. The smoking habit once again aided authors in their characterization. Women who smoked were no longer foreign sluts but sophisticated and independent. This forms an interesting contrast to the role allotted to male smokers, who now smoked to calm their nerves, which, presumably, the self-confident female smokers had inflamed. Men lit up in print at times of indecision, whereas women did so as a sign of self-assurance. The contrast is demonstrated by Agatha Runcible, It Girl extraordinaire in Evelyn Waugh's *Vile Bodies*, and Bertie Wooster, hero of P. G. Wodehouse's *Jeeves* novels. Agatha smokes Turkish cigarettes with spirit and nonchalance, ignores No Smoking signs, is utterly disdainful of serious matters, especially politics – for example, when she hears 'someone say something about an Independent Labour Party' she is 'furious that she had not been invited', and dies in a motor racing accident. Wooster, however, reaches for a cigarette whenever confronted by a domestic crisis – the choice of a tie, for instance. The cigarette had kept its trench association for men, even when transported into comic situations, but had developed entirely new ones in the case of women.

Meanwhile, in the United States, the cigarette had attained a similar position of eminence among tobacco habits, overtaking chewing tobacco and cigars as the nation's favourite weed delivery system. This rise in popularity was attributable, in part, to political causes. The American government had added another item to the growing list of reasons why to smoke. While its troops were away in Europe for World War I, various temperance movements had taken advantage of the wartime austerity measures to clamp down on bad habits among the godless. They struck first against alcohol and when the doughboys returned home they found their sacrifices rewarded with prohibition. The Volstead Act, rendering America dry, was ratified by Congress in 1919 as the eighteenth amendment of the constitution of the USA. Within months the first grand experiment in social engineering of the twentieth century had given birth to a new occupation – the bootlegger, or alcohol smuggler, and within a few years it had spawned the organized crime industry, whose spin-offs included national gambling and prostitution empires.

Fired by their success against the demon drink, the men and women of God set their sights on the weed. The evangelist Billy Sunday declared, 'Prohibition is won; now for tobacco.' The American people, aware that their pleasures could not be taken for granted, responded by smoking more than ever, especially cigarettes. In 1920 they consumed over 100 billion cigarettes for the first time. As had been the case in Great Britain, the American men were joined in smoking by their women who had been using the temperance movement as a Trojan horse to conceal their real aim of getting the vote. The influential *Atlantic Magazine* described the cigarette as 'the symbol of emancipation, the temporary substitute for the ballot'. When universal suffrage was granted and ratified by Congress as the nineteenth amendment in August 1920, female membership of temperance movements faded, and their consumption of cigarettes rocketed.

The veteran tabagophobe Miss Lucy Gaston fought a desperate rearguard action, even standing for president in 1921 against Warren Harding on an anti-cigarette platform. Harding won, perhaps on the strength of Thomas Edison's endorsement: 'Harding is alright. Any man who chews tobacco is alright.' Four years later the pious Miss Gaston died of throat cancer and must be assumed to have ascended to the company of her Maker, far above the smoke-filled earth and the flaming infernal regions awaiting users of the Devil's weed.

The science of etiquette, newly arrived in the United States, came to Miss Gaston's posthumous aid. Its text books considered women and smoking to be utterly incompatible, and, in a last ditch attempt to keep them apart, sought to restrict men smoking in front of them:

a gentleman should not smoke [in the presence of ladies] under the following circumstances:
When walking on the street with a lady.
When lifting his hat or bowing.
In a room, an office, or an elevator, when a lady enters.
In any short conversation where he is standing near, or talking with a lady.
If he is seated himself for a conversation with a lady on a veranda, in an hotel, in a private house, or anywhere where smoking is permitted, he first asks, 'Do you mind if I smoke?' And if she replies, 'Not at all' or 'Do, by all means,' it is then proper for him to do so. He should, however, take his cigar, pipe, or cigarette, out of his mouth while he is speaking. One who is very adroit can say a word or two without an

unpleasant grimace, but one should not talk with one's mouth either full of food or barricaded with tobacco.

In the country, a gentleman may walk with a lady and smoke at the same time – especially a pipe or cigarette. Why a cigar is less admissible is hard to determine, unless a pipe somehow belongs to the country. A gentleman in golf or country clothes with a pipe in his mouth and a dog at his heels suggests a picture fitting to the scene; while a cigar seems as out of place as a cutaway coat. A pipe on the street in a city, on the other hand, is less appropriate than a cigar in the country. In any event he will, of course, ask his companion's permission to smoke.

But not even manners could restrain the new woman. Like their British counterparts, American women had taken to dancing in public. A new form of music named jazz had arrived to stir their feet. It was instantly attacked by the fading temperance movement, which claimed jazz 'put the sin in syncopation'. The move lost the movement its remaining sympathizers. Women wanted to dance and smoke, and since they now voted, their desires could no longer be ignored.

American cigarette manufacturers showed less reserve than their British counterparts in targeting the female market, although they first carried out a little research into why women did smoke. To this end, they consulted Freud's American disciple, A. A. Brill, who attributed the desire to the familiar problem of 'lip eroticism', although he did not elaborate on why women should spend their lives in search of breasts. The simpler expedient of market research, i.e. asking women smokers why they smoked, came to different conclusions. What the girls wanted were brands they could call their own. But, packaging aside, in what way should a woman's cigarette be different from a man's? For example, should women's cigarettes be engineered to reflect perceived physical differences between the sexes? The manufacturers decided the answer was yes: women wanted lighter, cleaner,

prettier cigarettes. They wanted sugar 'n' spice and everything nice cigarettes. The firm of Philip Morris, which had
moved across the Atlantic and was now a New York niche
manufacturer, was first into the market. It revived one of
its English trademarks, named after the Duke of
Marlborough, which it abbreviated to 'Marlboro'. The new
brand – slogan: 'Mild as May' (1924) – began life as one
for the ladies, targeting 'decent, respectable' women. Initial
advertising copy adopted a softly softly approach – as if
there was still a need to justify women smoking: 'Has
smoking any more to do with a woman's morals than has
the color of her hair? . . . Women – when they smoke at all
– quickly develop discerning taste. That is why Marlboros
now ride in so many limousines, attend so many bridge
parties, and repose in so many handbags.' Marlboros were
not only branded but engineered for women, incorporating
an ivory tip of greaseproof paper to prevent them adhering
to lipstick. Marlboro's lead was followed by other manufacturers, some of whom had bolder opinions of women's
tastes. James Bull, a Milwaukee firm, ever conscious of the
delicate sex's weaker constitutions, introduced 'cocarettes'
'a soothing blend of refreshing Colombian coca-leaf and the
lightest Virginian tobacco, specially blended for the Lady's
need'.

In their rush to capture a share of this virgin market, the
US cigarette manufacturers did not forget their traditional
brands. By the middle of the 1920s all were in agreement
that advertising was the key to increased sales, and the
tobacco industry spent more on advertising than any other.
The US cigarette market was dominated by three big brands
– Camel, Chesterfield and Lucky Strike – who between them
accounted for 82 per cent of all sales in 1925. When any
one of these launched a new campaign or new publicity
stunt the others reciprocated with campaigns of their own.
Cigarette advertising was the proving ground for most of
the tricks of the trade, and there were few restrictions
governing the claims manufacturers might make for their
products.

Camel, the market leader in the early 1920s, was sold under the slogan 'Have a Camel'. This ambiguous advice was intended to imply that Camels were appropriate whatever the occasion. As the advertising copy elaborated, all the brand's customers 'have paid Camel the highest compliment . . . Camels never tire the taste'. American Tobacco responded by relaunching its Lucky Strike brand. Post-war America was a land of plenty – even of alcohol – drunkenness offences had risen threefold since prohibition. But the plenty was beginning to show itself on Americans' figures. American Tobacco decided to emphasize tobacco's well-documented property as an appetite suppressant to sell its cigarettes. 'Reach for a Lucky, instead of a sweet' was the slogan that launched its new campaign, which elicited a chorus of protest from US confectionery manufacturers. According to the Federal Food and Drugs Act, tobacco was not a food and hence it was fraudulent to claim a cigarette might serve as a substitute for a nourishing, properly labelled sweet. Luckies dropped the sweet reference but kept the central theme – that fat is bad and can be kept at bay by smoking. The American public agreed and Luckies' sales shot up. By 1930 it had overtaken Camel to become the market leader. Inspired by success in the fat field Luckies expanded laterally into health. If cigarettes could save people's figures then they must be good for them. They solicited the opinions of doctors across America – offering five free cartons of cigarettes to every one who answered the question right – and 20,679 physicians endorsed Luckies as a healthy brand. The survey results were incorporated within an advertising campaign showing people participating in outdoor pursuits, behind whom lurked a grossly overweight shadow of the sportsman/woman in question. In order to avoid such a transformation, the public were advised 'when tempted to over-indulge, reach for a Lucky instead'. Luckies also took advantage of the new fashion for sun tans, the first time this mania had surfaced in Western history since the days of the Roman emperor Claudius (10 BC–AD 54). Luckies' packaging had incorporated the slogan 'It's toasted' from its earliest days, a reference to a heat treatment process which was claimed to make Luckies a smoother smoke. This was now asserted to improve the cigarettes in the same manner as a sun tan did a healthy body: 'sunshine mellows, heat purifies . . .'

Tragically, as Luckies' campaign was peaking, excessive flesh ceased to be a danger for many Americans. The stock market collapse in 1929 was followed by the Great Depression, in which people lost their savings and their jobs. The three main cigarette brands responded to their customers' impoverishment by raising their prices. Despite this, and the continuance of insensitive advertising campaigns, cigarette sales flourished throughout the Depression. Although many smokers were reduced to getting their supplies second hand, i.e. by picking butts off the streets, people who could afford to buy cigarettes did, and many new people took up the habit. A new argument had arrived in favour of smoking, which did more for the cigarette than war, fat fear, votes or the dance halls had ever achieved.

Movies changed smoking for ever. The advent of the cinema allowed, for perhaps the first time in history, people separated by both distance and culture to enjoy identical visual experiences. The cigarettes, pipes and cigars that appeared giant sized in front of millions around the world made smoking look as natural as eating or kissing. Cinema effectively knocked out the need to justify smoking. Although this had diminished over the centuries, smokers still fabricated reasons for their habit, or took advantage of such myths that had developed in their societies. These certainly became weaker and less convincing in the twentieth century – few people bought a packet of Woodbines to ward off bubonic plague – but the excuses persisted until the advent of movies. Cinema presented tobacco habits shorn of justification. This was not intentional, but rather a handicap of the medium. Events on screen occur at a natural pace – the performance cannot be interrupted to allow the audience to question the actors, or, unlike a book, time to stop and reflect upon the passage of events. Justifying a tobacco habit would interfere with the plot of a film, and while reasons behind smoking might be implied, or mimed, they were seldom made explicit.

It was not, however, the mere depiction of the act that gave smoking such a boost, but rather who was smoking on screen that turned the people on. For the first time the population had public figures other than their rulers to adore. The cinema, like nothing before it, elevated the cult of appearance. Its idols moved, and after 1927, learned to speak and sing. Although only two-dimensional Pygmalions, they were so different, so much more convincing than the static representations people had hitherto respected, that their passions were inflamed. The actors and actresses played up to this adulation – some even saw it as their duty to live out their fans' fantasies in real life as well as on screen. Gloria Swanson, star of *Male and Female* and *Why Change your Wife?*, vowed 'I will be every inch and every moment a star.' Not only were movie stars idolized, they were imitated. And in the absence of an obvious physical resemblance, the best way of impersonating a hero or heroine was via their tobacco habit. People began to smoke because their favourite film star did. The cheapness and availability of cigarettes were to their advantage. The average audience member in the Great Depression (now global) could not afford the mansions, the yachts, the furs or the diamonds their idols enjoyed in flickering black and white, but they could buy the cigarettes and so share a portion of the dream. Smoking was an aspiration everybody could fulfil.

Stars were subjected to an unprecedented level of public scrutiny. They were examined more deeply than the politicians and sporting heroes who had hitherto commanded the people's esteem. Such curiosity resulted from the intimacy inherent in celluloid. Whereas the old idols had gained their reputations by saving nations or winning games, the new had acquired theirs by looking irresistible in pyjamas or by acting out the most commonplace events, including doing the washing up. Audiences felt that they knew so much already about their screen gods and goddesses that they needed to know the rest. What brand of cigarettes did they smoke, for instance? Cigarette manufacturers were quick to spot the advantages these new heroes could give their brands. Lucky Strike, followed by its competitors, assisted fans by publishing the names of thespians who smoked their brands and by paying them to do so. They signed on Al Jolson, the blacked-up white Russian star of *The Jazz Singer* (1927), and also recruited young Frederick Austerlitz, better known by his stage name – Fred Astaire.

The film industry, meanwhile, had settled in a new town on the west coast of the USA, the fame of whose name now rivals Mecca. Hollywood's dry, stable climate, and its collateral advantage of cheap electricity, made it the ideal location for movie-making. Hollywood made the world's dreams, but made them in America's likeness. Hollywood was a sober, god-fearing town with the power to outlaw lewd conduct in the films made within its jurisdiction. Fear of censorship prevented directors from portraying even kissing on screen to a planet whose population was exploding. The silent movies of pre-1927 relied heavily on gesture and the cigarette became not just a prop, but an allegory. If a celluloid heroine felt desire, she was filmed smoking or asking for a match. Early female stars such as Bette Davis appeared to live on smoke. A cigarette served as an obvious and potent metaphor for forbidden carnal activity. This role as a lovers' go-between was strengthened when the behaviour of stars off screen brought the threat of censorship closer. Tragically, some Hollywood idols had personal lives very different to their screen personae. Conscious of the influence their players wielded, and after some had succumbed publicly to out of character afflictions (Fatty Arbuckle, child sex, Mary Pickford, divorce, Wallace Reid, a drug overdose), morality clauses were inserted into actors' contracts and the industry, pre-empting the temperance brigade, adopted its own code of conduct, which imposed strict restrictions on celluloid encounters between men and women. Smoking was one of the few joint activities they could be depicted enjoying.

Why smoke?

This new convention – the cigarette as a substitute for a kiss, or more – captured audiences' imaginations all over the world. Just as tobacco had once served as money in the United States, it now acted as the international currency of desire. Most stars with a reputation for sexuality would have at least one still photograph in their publicity portfolio showing them smoking.

Although cigarettes were the principal beneficiary of

Hollywood, other tobacco habits were also represented on screen in accordance with the prevailing conventions of literature. What a character smoked on film was a visual clue to their personality and background. The Victorian hierarchy of taste continued to be applied. A pipe smoker was a thinker, or a dependable member of the middle class. Fritz Lang's classic *M* contains the most extreme visual interpretation of such men drawing inspiration from their pipes. As detectives and town councillors ponder the identity of a murderer, a monstrous fug develops in the room – as if to emphasize the intensity of their thought. Cigars were a celluloid power symbol – or penis substitute, according to Freud's now estranged disciple Carl Jung. Businessmen, those involved in the entertainment industry and gangsters were shown smoking cigars. Edward G. Robinson, prince of screen hoodlums, could bite the tip off his cigar in a manner so threatening that female audience members fainted.

Celluloid exposure resulted in a boost for the quality end of the cigar market. Cheap cigars had largely been superseded by cigarettes as everyday smokes, although they were still bought for special occasions, or to celebrate an event, like the birth of a child. Jung interpreted this 'tradition' sexually, only to receive a rare rebuke from Freud – 'sometimes a cigar is only a cigar'. The cigar also carved a film niche for itself in comedy, where it was carried like a banner by Groucho Marx. Groucho, like that other Marx of the previous century, smoked cigars in real life and used them as an extension of his character. His cigars played the role of ventriloquist's dummy – themselves a party to conversation as well as a prop. On one occasion, when a woman told him she had twenty-two children because she loved her husband, Marx responded, 'I like my cigar, too. But I take it out of my mouth once in a while.'

Not every member of humanity had joined in the rush towards egalitarianism by smoking cigarettes, and the cigar became the badge of those who stood apart. Often the creators of the literary and film fantasies fed to the common man favoured cigars over the cigarettes they peddled in their oeuvres. While the film stars smoked cigarettes, to keep in touch with their fans, or because they were paid to, film directors smoked cigars. Figures such as Darryl F. Zanuck ('I would rather die with a cigar in my mouth than boots on my feet'), who had started smoking cigars to appear older when he became head of production at Warner Brothers in 1925 at the age of twenty-three, helped to create the stereotype of the director with an oversized Havana in his mouth. His example was followed by Sir Alfred Hitchcock and Orson Welles. The writers who portrayed the rise of the cigarette also preferred cigars. Evelyn Waugh claimed he only wished to earn money enough in his lifetime to be able to drink a good bottle of wine every day and smoke a Havana cigar. His sentiments were echoed by Somerset Maugham, who observed: 'when I was young and very poor . . . I determined that if I ever had any money I would smoke a cigar – every day after luncheon and after dinner'. When he was very old and very rich, he did so. P. G. Wodehouse, the kindest chronicler of the 1920s and 1930s, whose characters reach for a 'gasper' in times of stress, smoked Cuban cigars exclusively.

The influence of celluloid on smoking was not limited to the portrayal of fiction. Cinemas showed news and documentaries in addition to feature films. The public were able to see their rulers with more detail than their representations on their national currencies. Not only film stars in fictional roles but people playing themselves in real life could be viewed and judged. The tendency of men in power to smoke cigars was noted and this habit too became aspirational.

And so tobacco continued its triumphal march. By 1930, as Count Corti, a tobacco historian, observed, most people in industrialized societies smoked: 'a glance at the statistics proves convincingly that the non-smokers are a feeble and ever dwindling minority. The hopeless nature of their struggle becomes plain when we remember that all countries, whatever their form of government, now encourage and facilitate the passion for smoking in every conceivable way, merely for the sake of the revenue which it produces . . .' Not only was the world full of smokers, but they could enjoy their habit free of criticism. Whether rich or poor, famous or obscure, they had one common habit. It was observed of cigarettes that they were 'the most democratic commodity in common use'. At last neither sex nor age was a barrier to smoking. In 1934, emblematic of the progress smoking had made in the jazz age, the elegant southern heiress Eleanor Roosevelt became the first First Lady of the United States of America to smoke a cigar in public.

14

Verboten

*Smoking suffers under the Nazis in
Germany – tobacco at war around the
world – tobacco in prison camps and on
the black market – cigarette manufacturers
and Hollywood contribute to the war effort*

Throughout the 1920s and 1930s, the world was a smoker's paradise. Manufacturers competed for their custom with a cornucopia of products, taxation was light, if omnipresent, and smokers were part of an international fellowship with its own language and customs, most of which related to love and friendship. This unusual period of calm in tobacco's troubled history did not last. At the same time as Freud was preaching his theories in Vienna, a young, impoverished artist was breathing the same air and formulating his own doctrine for humanity. Adolf Hitler was too poor to smoke in his Vienna days, and in any event was opposed to the habit after being punished for an illicit cigarette while at school.

When Adolf traded art for politics he found the success that had eluded him among his paint pots and badger hair brushes. After declaring himself Führer of Germany's Third Reich, Adolf's preferences became the policies of his Nazi party. Among these was the first scientific anti-smoking initiative. Hitler's Nazis compiled all available information on tobacco and health and searched this record for crimes against humanity. While its scientists were busy with their slide rules and Petri dishes the government commenced a war against smoking, attacking the habit with propaganda and taxes. It began with a poster campaign, which featured a smoker's head being crushed under a jackbooted heel. Hitler supported his pet crusade with rhetoric: tobacco was the 'wrath of the Red Man against the White Man for having been given hard liquor'. Hitler was particularly opposed to cigarettes appearing between the lips of the fairer sex. Members of the Nazi Women's League forswore all tobacco as part of the process by which they dedicated themselves to the propagation of the master race. Hitler was also happy to serve as a role model for the non-smoker. The cover of *Auf der Wacht* in 1937 featured a portrait of the dictator in a pensive mood with the caption, 'Our Führer Adolf Hitler drinks no alcohol and does not smoke . . . his performance at work is incredible.'

Germany's cigarette manufacturers decided the best form of resistance to the onslaught was collaboration, and donated large sums to the Nazi party. Some made pathetic attempts at appeasement, including the market leader, Cigaretten Bildendienst, which offered coupons with its cigarettes that could be exchanged for a coffee-table book on Hitler. They seem to have succeeded for surprisingly, in the authoritarian atmosphere of Nazi Germany, where failure to conform to the Führer's wishes was persecuted with intensifying rigour, smoking increased between 1932 and 1939 from 570 cigarettes per capita per annum to 900. The tax revenues generated by smokers were significant, exceeding 1 billion Reichsmarks in 1938. Although officially tabagophobe, Hitler's Germany was the world's largest tobacco importer, taking approximately 100,000 tons of the weed annually.

At the end of the 1930s the anti-tobacco programme was expanded to include bans on smoking in public places and on transportation, outright bans on smoking among pregnant women and members of the Luftwaffe. Berlin even renewed its municipal ban on smoking out of doors that had led to revolution a century before. Taxation was raised to punitive levels (for that age – they were less, for instance, than those currently imposed by the British government) and by 1941 tobacco taxes were responsible for one-twelfth of all government income. These draconian measures, and intervening events, at last had some effect on German smokers, whose habit declined from 1939 onwards.

Hitler's scientists meanwhile had come up with some disturbing theories about smoking. In 1939, Franz H. Müller was the first to use case-control epidemiological methods to document a relationship between smoking and lung cancer. Müller concluded that the 'extraordinary rise in tobacco use' was 'the single most important cause of the rising incidence of lung cancer'. Müller's work was followed up four years later by Eberhard Schairer and Eric Schöniger who reached the same conclusions.

However, the credibility of the scientific campaign against smoking suffered by being part of a wider movement intended to establish racial purity. The Nazi party's aims included global improvement as much as self-improvement, and the Third Reich's attempt to enforce its policy of 'Liebensraum', or 'living space', which was intended to secure land for the perfect Aryans it hoped to breed, launched the series of conflicts that led to World War II.

World War II was truly global. The remotest Pacific atolls were carpet bombed with high explosives, while submarines and convoys played cat and mouse amidst the Arctic ice. As had been the case in World War I, tobacco use flourished and cigarettes spread to the last, most distant places on the planet. In countries where cigarette smoking was established, like Great Britain, cigarettes were accepted as an absolute necessity in hard times and did not join such essentials as bread, milk and beef on the ration list. Cigarettes were deemed essential for the maintenance of British backbone at home. Consumption amongst British women rose sharply, especially in areas subjected to Nazi bombing: 'In London during the heavy raids I found smoking a great help. My consumption has gone up over 100 per cent. It started in

'You don't smoke it – it smokes you!' Nazi anti-tobacco propaganda

September – during the Blitz. I found smoking kept me from
getting jittery.'

As in World War I, troops of all the major combatant
countries were issued tobacco. Cigarettes were provided free
to British soldiers as part of their rations. The 'compo-
ration', which was the main source of food for British and
Commonwealth troops on most fronts, provided seven cigarettes per man per day. These could be supplemented by
purchase at NAAFIs, and were further augmented by the
comfort parcels sent by relatives and well-wishers in the UK
and Commonwealth. Once again, officers purchased extra
cigarettes for their own men before going into action. Even
General Montgomery (an ardent anti-smoker) personally
supervised the distribution of cigarette cartons prior to the
battle of el Alamein. Montgomery's opposition to smoking
contrasted sharply with Winston Churchill's infatuation with
cigars. The difference between a general's and a prime

minister's approach to conflict is illustrated by the following exchange between them. When Montgomery told Churchill, 'I do not drink. I do not smoke. I sleep a great deal. That is why I am 100 per cent fit', Churchill responded, 'I drink a great deal. I sleep little, and I smoke cigar after cigar. That is why I am 200 per cent fit.'

In the United States, President Roosevelt declared tobacco an essential wartime material and granted military exemptions to those who grew it. American troops received a more generous tobacco allowance than their allies of between five and seven packs of cigarettes each per week. From 1944 onwards, cigarettes were incorporated in the US 10-in-1 combat rations. The US government further encouraged soldiers' smoking habits by the introduction of slogans such as 'Smoke 'em if you got 'em' and the incorporation of smoking breaks into marches. The cigarette manufacturers made a virtue of their product's necessity, and ran adverts emphasizing the cigarette's role in the lives of the men who were fighting to save the world from totalitarianism. For example, in 1942, Lucky Strike faced a shortage of the green dye used on its pack which was also needed for camouflage uniforms. Luckies turned this to their advantage, changing the pack colour from green to white and selling the new one with the slogan: 'Lucky Strike green has gone to war.'

Thanks to their Führer's concern for their well-being, German troops did worse than their fathers had in 1914. Hitler had ordered that tobacco rations be distributed to soldiers 'in a manner that would dissuade' them from excessive smoking. The wartime ration was six cigarettes per head per day. Extra supplies of up to fifty cigarettes per month might be purchased, but these were taxed at levels approaching 90 per cent of the retail price. The elite SS soldiers fared a little better, being provided with their own-label brand of 'Sturmzigaretten' ('Storm-cigarettes'), made by the Sturm-Abteilung. The German's Axis allies in the Imperial Japanese army were also supplied with cigarettes by their government. Their favourite brand was a Chinese

one named 'Ko-ah', which translates literally as 'Golden
Peace'. Ko-ahs had been in plentiful supply ever since the
invasion of China.

Russian tobacco rations were distributed less clinically.
Each Red Army soldier was allotted a daily handful of rough
Makhorka tobacco and (in theory) 100 grams of vodka.
Russian soldiers always rolled the tobacco in newspaper as
the newsprint was commonly held to enhance its taste.
Russian troops smoked constantly in battle. As an anti-tank
rifleman speaking to Konstantin Simenov observed (1942),
'It is permissible to smoke in battle, what is not permissible
is to miss your target, miss it just once and you will never
light up again.'

Cigarettes fulfilled a less glamorous function on the battlefields than they had in the years prior to the war. Soldiers
did not smoke to appear sophisticated. They used tobacco
for the same reasons in the 1940s as the musketeers, knights
in armour and pikemen of the Thirty Years War three
centuries before them. Those who did not smoke before the
war took the habit up – a GI in Normandy recalled talking
to a fresh recruit who 'never smoked cigarettes until he
landed in France on D-Day but after that he smoked one
after another. He was about the tenth soldier who had told
me the same thing. A guy in war has to have some outlet
for his nerves and I guess smoking is as good as anything.'
A non-smoking colonel in the British 2/5th Queen's Regiment
came to the same conclusion. He 'had at one time tried to
ban smoking during exercises, but discovering it produced
an appalling effect on morale, he determined that the troops
would always have a liberal supply . . .'. Before his troops
commenced their next action 'He bought five thousand and
loaded them into his jeep.'

Soldiers were intensely brand sensitive. The British
preferred Woodbines, Players and Gold Flake, the Americans
Camels, Luckies and Chesterfields. Regrettably, the manufacturers' war efforts could not keep up with their tastes
and the cigarettes in ration packs were often government

made. Among other horrors forced upon Commonwealth front line troops in their combat rations were 'Cape to Cairo' (made in Egypt), 'V', made in India, and 'RAF' made in England out of the cigarette butts swept up from cinema floors. Clearly, the pictures on the packets were important to the troops. They were a bond to their homes and past lives as civilians. Cigarettes were almost unique among the disposable items soldiers carried into battlefields, in that they used the same branded product every day in peacetime too. This is why brands were precious on active service, and why state-made substitutes did not satisfy. Uncensored complaints appeared in the press: 'a man who is suffering the hardships and privations of the Front Line should be given the best [cigarettes] his country can give'.

An understanding of the reasons why soldiers smoked is assisted by investigating the reasons why they were supplied cigarettes. Military training was intended to achieve total control of soldiers, so that even their bodies were 'mere implements of warfare'. Soldiers were provided with the sustenance necessary to achieve this state, and no more. As fighting machines, they were fed, watered and rested, but seldom indulged. Armies would not have supplied their soldiers with cigarettes with the same dedication as food and ammunition if they had been considered injurious or inappropriate. In some cases, such as in the Burma campaign, cigarettes were flown in by aircraft and accorded the same priority as medical supplies. It was accepted that smoking increased a soldier's efficiency. It achieved this, in part, by steadying his nerves: 'When he is mortared or shelled he is deathly afraid . . . and chain smokes, curses or prays, all of this lying on his belly with his hands under his chest to lessen the pain from the blast.'

The faint reputation cigarettes had attracted as being injurious to health died a death on the battlefields. Soldiers were painfully aware of the importance of health and longevity, and did not take up habits that might prejudice either. Their existence had enough risks and too few pleasures. The extreme demands on their bodies did not encourage them to undertake further punishment, although the perils of smoking were sometimes the subject of a good deal of irony, as this exchange between two Cumbrian soldiers of the 14th Army on the Burma campaign illustrates:

'Smoking boogers yer wind' conceded Wattie, inhaling with satisfaction . . .
'Knackers' retorted Grandarse . . . 'Fags nivver done me nae harm. Ah joost smoke it in and fart it out, an Ah's in grand fettle.' He demonstrated thunderously, guffawing, and those nearest recoiled in disgust.

From the soldier's point of view, cigarettes were more than mild anaesthetics. Cigarettes formed an umbilical chord linking soldier to civilization. There was little else in the daily grind of being bombed, burned and maimed, of killing or being killed in foreign countries to remind them of home. Even trying to conceive of what they were suffering in lay man's terms was impossible. Here is an attempt to describe the rigours of war in civilian language by a fighting soldier:

Dig a hole in your backyard while it is raining. Sit in the hole while the water climbs up around your ankles. Pour cold mud down your shirt collar. Sit there for forty-eight hours, and, so there is no danger of your dozing off, imagine that a guy is sneaking around waiting for a chance to club you on your head or set your house on fire. Get out of the hole, fill a suitcase full of rocks, pick it up, put a shotgun in your other hand, and walk on the muddiest road you can find. Fall flat on your face every few minutes, as you imagine big meteors streaking down to sock you . . . snoop around until you find a bull. Try to figure out a way to sneak around him without letting him see you. When he does see you, run like hell all the way to your hole in the back yard, drop the suitcase and shotgun, and get in. If you repeat this performance every three days

for several months you may begin to understand why
an infantryman gets out of breath. But you still won't
understand how he feels when things get tough.

Comprehensive surveys carried out by the United States on
its returning troops concluded that when things did get tough
soldiers were afraid, mildly religious, and saw the purpose
for their presence in the line of battle not so much as the
consequence of patriotism, or belief in ideals, but rather in
terms of loyalty to their immediate comrades. Certainly cigarette smoking was part of the bonding process between
troops at the platoon level, where they were invariably
shared in moments of respite. Even on the Russian front
there are many touching examples of the power of smoking
to soften the horrors of war. Some of the most poignant of
these come from the siege of Stalingrad when the German
6th Army was surrounded after a daring Russian counter-
attack. With scant provisions, Christmas must have been
particularly bleak for the encircled German troops and cigarettes are a recurring theme in letters home. One lieutenant
described making a gift of the last of his cigarettes to his
men: 'I myself had nothing and yet it was one of the most
beautiful Christmases and I will never forget it.' Of more
than 250,000 troops encircled in the battle for Stalingrad,
only 25,000 returned behind German lines.

An old Amerindian and more recent Virginian use for
tobacco was revived in the POW camps of World War II,
where cigarettes became the unofficial currency. Cigarettes
also served as an interface between prisoner and guard. The
guards could amuse themselves watching prisoners scramble
for cigarettes, but were also happy to accept them as bribes.
They were important human links between people of
opposing creeds, and sometimes alleviated the suffering of
the prisoners. The following description of conditions in
Stalag 4B, made by George Rosie of the US 106th Airborne,
provides an accurate summary of tobacco's pre-eminence in
prison camp affairs:

Of the four camps George had been in, Stalag 4B was
by far the best run camp. The Brits had everything
organized. The Sergeant Major let the men know in no
uncertain terms that he was in charge of the barracks.
The Red Cross parcels and the food could be bartered
at their exchange store; the money was cigarettes, and
every parcel had a cigarette price . . . There were six
Americans captured at Dunkirk who were in the
Canadian army. Canadian people would get names of
POW's from their newspapers and send them cartons
of cigarettes. Because these Canadians had been in
Stalag 4B for so long, they were getting as many as
twelve cartons of cigarettes a month, which made them
rich men in POW camp economy.

Cigarettes also served as a currency to a much wider extent
in Germany post-1945, where prisoner-of-war camp conditions were repeated on a national scale. The collapse in
German smoking Hitler had wished for had happened at
last. The German tobacco industry could only operate at 10
per cent of its pre-war capacity and imports were at the
whim of the occupying Allied forces. It was possible to buy
the young German women who had been prohibited from
smoking to ensure the future health of the master race for
a few cigarettes. Everything had a cigarette price. They were
perfect currency substitutes – easy to transport and store,
of uniform branded quality, and easily divisible – a bargain
could be struck for one, ten or 10,000 cigarettes. When
Germany was partitioned families moving from east to west
carried their wealth in the form of cigarettes, which could
be broken out in ones or twos to secure small favours or
buy a drink, or could be handed over in cartons for larger
purchases or desires. Cigarettes also kept their value. Like
bullets and wedding dresses, they fall into the category of
things that are only used once. Even when sizeable cigarette consignments reached Germany, these were soon
consumed. Steady demand and inconstant supply therefore
supported cigarettes' monetary value, and they had the

added advantage that they could be smoked. Their superiority to Reichsmarks was demonstrated by the fact that the one was used to light the other.

Tobacco's status as currency in post-war Germany made it the perfect cash crop. 'We too grew tobacco, although nobody in our family smoked. But money was worthless, and you could get everything for tobacco. We had some land, and cultivated wheat which we had to cede to the state . . . but right in the middle of the wheat, where you could not see it, we planted tobacco.' Germans even grew tobacco in their window boxes to gain access to some kind of hard currency. The government allowed tobacco cultivation, but only twenty-five plants could be grown tax free, and no more than 200 per individual in total. The infamous tobacco police of the previous century reappeared to enforce these restrictions. Home-grown, however, fetched far less on the black market than smuggled Allied cigarettes. Cigarettes containing home-grown tobacco were nicknamed 'lung torpedoes' – an indication that some of the Nazis' scientific research had struck home. The men in the Nazi administration behind the anti-smoking campaign did not survive to see the belated triumph of their policies. Health führer Leonardo Conti committed suicide in 1945 while in prison awaiting prosecution for his part in the Nazis' euthanasia programme. Gauleiter Fritz Sauckel, the guiding light of the anti-tobacco institute, was executed in 1946 for crimes against humanity. Count Corti's final assertion in his *History of Smoking*, written nearly a decade before the war had broken out, appeared to be a prophecy fulfilled: 'If we consider how in the past the efforts of the most absolute despots the world has ever seen were powerless to stop the spread of smoking, we may rest assured that any such attempts today, when the habit has grown to such gigantic dimensions, can only result in a miserable fiasco.'

Tobacco manufacturers in the Allied countries, in particular in the United States and Great Britain, had had a good war. Weapons and destruction had provided American tobacco companies with a rich source of imagery for their domestic advertising campaigns. Pall Mall, the first king-sized brand of cigarette, used advances in weaponry to support its own claims of technical superiority. Adverts with pictures of anti-tank guns and armoured cars were supported with copy such as the following:

> In cigarettes, as in armored scout cars, it's modern design that makes the big difference!

> Fast as a racer, staunch as a tank! No wonder these new streamlined scout cars are a vital development in mechanized warfare. Modern design makes the big difference – a difference that's mighty important in national defense.

> Talk with the men who ride these cars. They'll tell you that in cigarettes, as in armored scout cars, it's modern design that makes the big difference.

> Now, at last – thanks to modern design – a truly fine cigarette provides in fact what other cigarettes claim in theory – a smoother, less irritating smoke – Pall Mall.

RJR's Camel brand took a different approach, running adverts that showed its cigarettes involved at the point of conflict. Its copy was directed at non-combatant Americans, whom it urged to smoke more:

> TIN FISH – that means torpedo in submarine language. The phrase, 'the smoking lamp is lit' means Camels are in order – for men in the Navy, the favorite cigarette is Camel.

> HIDE-AND-SEEK. A deadly game of it with the T.N.T. of depth charge and torpedo. That's a game only for steady nerves. But, then, what isn't in these days – with

all of us fighting, working, living at the highest tempo in years. Yes, and smoking, too – perhaps even more than you used to.

If Camels are not your present brand, try them. Not just because they're the favorite in the service or at home – but for the sake of your own smoking enjoyment, try Camels.

The depth of association American cigarette brands had achieved with the country's troops by VE Day was reflected in the names allotted to the French transit camps where GIs waited before being shipped back home. The camps at Le Havre and Marseilles were named after the soldiers' favourite brands: Camp Chesterfield, Camp Lucky Strike, Camp Pall Mall and Camp Philip Morris. The intention was that veteran soldiers would feel comfortable staying in a camp with a familiar American icon for its name. Camps in other regions used the names of US cities, such as Camp New York and Camp Atlanta. Evidently, soldiers were believed to associate cigarette brands with the country they had been fighting for as strongly as the names of its principal cities.

Manufacturers also enjoyed a hugely increased customer base, thanks to the number of people who had taken up smoking during the war. Cigarettes had enjoyed a quasi-official status – they had been part of soldiers' rations, and in this sense, part of the war effort itself. Outside of Germany, no one had questioned their potential to harm humans. Not only cigarettes but other tobacco products shared the glory. All Allied leaders had smoked – usually conspicuously. General Douglas MacArthur returned, with a pipe. Winston Churchill gave his famous V for victory symbol balancing a cigar between his splayed fingers. Stalin smoked a pipe which he filled with tobacco from torn up Russian cigarettes.

Tobacco also took part in the peace process. Just as the Cherokee had kept two pairs of pipes, one for war, one for peace, tobacco was offered from victors to vanquished as a

MacArthur returns

token of consolation, and symbol of the potential for new friendship. General MacArthur used a cigarette in this role when he received the Emperor of Japan a few day's after his country's formal surrender: 'I offered him an American cigarette which he took with thanks. I noticed how his hands shook as I lighted it for him. I tried to make it as easy for him as I could, but I knew how deep and dreadful must be his agony of humiliation.' The Americans, in particular, were generous to their former enemies. As part of the Marshall

Plan, which paid for Europe's reconstruction, the US government sent a gift of 210 million cigarettes to the German
authorities, to be used as incentives for postal and railway
workers.

Hollywood had also had a good war. The Commonwealth
and American armies were catered for by entertainments
associations and on some campaigns were treated to films,
particularly towards the end of hostilities. The most significant of Tinseltown's early war movies was *Casablanca*
(1942), which opens with Humphrey Bogart's hand signing
a cheque, then moving to collect a smouldering cigarette
from a glass ashtray. *Casablanca* was an allegory that packed
all the themes of the war in to a Moroccan nightclub, and
set out the reasons why the United States could not stay
neutral in Europe. It was not possible to sit back for ever
and claim 'I stick my neck out for no one.' Bogart became
the quintessential screen cigarette smoker – brave, under-
stated, cynical, yet strong, who used his cigarettes to disguise
his feelings rather than reveal them. Hollywood soon found
him a queen in Lauren Bacall, whose first line in *To Have
and Have Not* (1944) – 'Anybody got a match?' – is followed
by a travelling shot of the trajectory of a matchbox, thrown
by the king himself.
   Film-makers, like painters before them, had discovered
the decorative effect of smoke during the war years. They
had also expanded on the repertoire of smoking types they
had inherited from literature. The way a man or woman
handled a cigarette on screen betrayed not only their background, but also their state of mind. Fingers could shake,
matches spill to portray nerves, a cigarette could be lit effort-
lessly, flamboyantly (literally), sexily, intriguingly – the possi-
bilities were numerous, and Hollywood exploited them all.
When the war ended, it set about glorifying the conflict.
Some of its biggest stars traded their buckskins for khakis
and were shown dealing out frontier justice to members of
the Axis forces. The cigarette is omnipresent in such movies,
especially in battle scenes. One of the most popular of this

genre, *The Sands of Iwo Jima* (1949), features John Wayne
and a supporting cast of US marines who spend their time
on screen acting out the slaughter of unfeasibly large hordes
of the yellow peril. When the Japanese Imperial Army is
finally seen off, Wayne drawls, 'I never felt so good in my
life. How about a cigarette?'

15

The Spy Who Loved Me

*Tobacco in the Cold War – teenagers and
smoking – Kool and the gang: filters and
mentholated cigarettes – the appearance of
'healthy' cigarettes – tobacco on television
– Britain's 'angry young men' – tobacco on
Her Majesty's Secret Service*

The Russians did not join the Americans in giving splendid
gifts of cigarettes to the Germans. They no longer saw eye
to eye with their wartime partners in many other matters,
and the Allies became enemies in an undeclared conflict.
They closed their borders to one another's citizens, goods
and cultures, and as Winston Churchill observed in 1946:
'From Stettin in the Baltic to Trieste in the Adriatic, an iron
curtain has descended across the continent of Europe.' The
resulting stand-off was termed the 'Cold War', whose battles
were fought indirectly, on the territory of weak, newly post-
colonial or neutral countries. Following the proliferation of
nuclear weapons, none of the principal protagonists wanted
to risk their own homelands being consumed in flames and
radioactive fallout. Casualty lists from Hiroshima and
Nagasaki had continued to mount for a decade after the
atomic explosions that had destroyed them, and the realiz-
ation that an atom bomb meant more than a giant blast had
sunk into the global conscience.

Whereas the Axis had been shattered by warfare, the

Allies' friendship was ended by political divergence. While they had been united in how they should fight the Germans, they differed in opinion as to how to treat their own citizens. The Russians subscribed to communism, a system of government under which everybody owned everything, but no one had anything of their own. By way of contrast, nobody in the capitalist West owned everything, but most people had something they could call their own. And since the capitalist states were enjoying what they termed a 'boom' in the 1950s, more and more capitalists owned more and more things. Even their children could afford to go out dancing. This time, instead of memorizing footwork, they swung their hips to a music developed from the slave chants of the Mississippi delta, and 'rock 'n' roll' was born. Rock 'n' roll was a simpler, more strident type of music than that which hitherto had been popular in the West. Opinions as to its merits were divided. It drove a wedge between generations. Old people found rock 'n' roll raucous and repetitive. They did not understand its lyrics, nor how it could have the power of enchantment. The young, in contrast, loved rock 'n' roll. The gap that developed between generations resulted in the birth of a new social grouping, named teenagers, whose idols were rock 'n' roll stars.

Part of the appeal of rock 'n' roll was its aura of rebellion. Its songs and its icons promoted the enjoyments that the parents of teenagers sought to curtail, such as premarital sex, drinking and smoking. Many of the early rock 'n' roll stars were smokers, including Chuck Berry, Buddy Holly, Elvis Presley and Little Richard, all of whom had experimented with smoking when teenaged delinquents. Jerry Lee Lewis, another smoker, exploiting, perhaps, the symbolic link between cigarettes and flames, became the first rock 'n' roller to light up not only a cigarette, but his instrument on stage, when he set fire to his piano at a concert in 1958: 'They still talk of that show, how Jerry Lee had the crowd screaming and rushing the stage, how he took a Coke bottle of petrol from his jacket pocket and doused the piano with

one hand as the other hand banged out "Whole Lot of
Shakin' Going On", how he set the piano aflame, his hands
still riding the keys like a madman as the kids went finely
and wholly berserk with the frenzy of it.'

These social developments were observed by America's
tobacco companies with the keenest interest. The new
teenage market displayed attractively tribal attributes, which
the cigarette manufacturers longed to exploit. Teenagers
were wonderfully imitative – they mimicked their icons with
a devotion and attention to detail far exceeding that of
adults. But the question was: should they have their own
brands, such as had been devised for their mothers and
grandmothers? After rebelling against their parents would
they also reject their smokes? It seems the answer was no.
Teenagers wanted to be treated as grown-ups. They were
old enough to marry, fight wars, and to work for a living.
They hated adults for their condescension, not their cars or
cigarettes.

Happily, the first great teen icon of the silver screen was
a smoker, and a stylish one at that. James Dean, star of
Rebel without a Cause (1955), the definitive teen angst
movie, in which Dean uttered such seminal phrases as
'You're tearing me apart', went straight to the hearts of
America's youth. Dean smoked conventional adult brands
on screen and in real life – a matter of relief to the American
tobacco companies, as it meant that they did not have to
devise new cigarettes for adolescents. They were already
having changes forced upon them by the market.

Competition for the ever diminishing band of non-
smokers had increased. It was no longer realistic for tobacco
companies to base cigarette sales projections on converting
innocents to the weed. Once everyone smoked, short of
encouraging them to smoke more or to breed, the tobacco
manufacturers would run out of new customers. Although
some industry figures viewed this prospect with pleasure,
believing that when it occurred they could raise their prices,
others contemplated its inherent limitations with gloom.
When fresh smokers ran out, there was going to be trouble.

Rivalry was already intense – any advantage a new advert or new gimmick achieved for one brand resulted in it being imitated. When Lucky Strike claimed its cigarettes were healthier than others', rival manufacturers responded with counter-claims for their own brands. Health turned out to be the wrong issue to raise in cigarette advertising, for it implied that some cigarettes, in being less healthy, were unhealthy.

A niche market of healthy smokes, for invalids and such like, already existed, for which special products had been developed, including mentholated cigarettes. Mentholated cigarettes were impregnated with an extract of peppermint, that was also used by vets as a local anaesthetic. The first menthol brand was the evocatively named 'Spud', which had been introduced in the 1930s. It was not alone for long. A second tier manufacturer named Brown & Williamson, sensing a future in healthier cigarettes, introduced its own menthol brand which it baptized 'Kool'. Kool's marketing promised smokers the opportunity to indulge without fear of laryngeal irritation: 'Give your throat a Kool vacation!' By the 1950s there was a range of menthol brands and the category was an independant and profitable market.

A second type of healthy cigarette, the filter tip, which had a small section of wadding attached to one end had also been in existence for some years. Filters were targeted at hygenic smokers, as they prevented tobacco shreds from sticking to smokers' teeth. The leading filtered brand was Viceroy, Kool's stablemate, whose filter used a roll of cellulose acetate. As Viceroy's sales climbed in the 1950s, other manufacturers took notice of its unique selling point. Lorillard, one of the majors, decided filters were the future and harnessed the most advanced technology of the day in designing one of its own for a new brand. Exhaustive studies were made of the thermodynamics and chemistry of cigarette combustion. Lorillard wanted to know exactly what was going on inside the little paper wrapper. How hot was a burning cigarette? What did its smoke consist of? Did it have any dangerous ingredients, and if so could these be separated out before they reached the smoker?

After finding some undesirable by-products of combustion in tobacco smoke Lorillard accepted the utility of a filter, and resolved that the most practical kind, short of attaching a miniature chemical plant to the end of each cigarette, was a physical barrier, and that the best material for this was asbestos. Asbestos was a popular material in the 1950s. Its fibrous structure made up of ultra-fine, barbed filaments made it an excellent heat retardant and filter. In those days, the discovery of a new asbestos mine was announced in the newspapers and celebrated. Lorillard slotted an asbestos section into their new filter which they branded 'micronite', and the cigarette they attached it to was baptized 'Kent'. The ensemble was promoted as the first sanitary cigarette. 'Do you like a good smoke but not what smoking does to you?' Kent's adverts inquired.

As the 1950s progressed, the health issue became increasingly important in cigarette advertising. Although none of the other majors rushed after Lorillard into filters, some sought to counter the implication that their brands were dangerous by setting up a smokescreen. Liggett & Myers, the manufacturers of Chesterfield, bombarded the airwaves with an advertising campaign based on a 'scientific' survey it had sponsored that proved its brand to be entirely harmless. Radio was still the principal broadcasting medium, and its presenters were most partial towards their advertisers. Arthur Godfrey, a star of his age, assured his listeners that Chesterfield's research 'ought to make you feel better if you've had any worries at all about it. I never did. I smoked two or three packs of these things every day – I feel pretty good . . . I never did believe they did you any harm and we – we've got the proof.' Freud would have made much of the use of 'we' by a presenter whose audience presumed he was his own man. Even in the cavalier spirit of the age Chesterfields were judged by the American authorities as having gone too far and the Federal Trade Commission took out an injunction to prevent them going further. Liggett & Myers responded by introducing their own brand of filter tips which they christened 'L & M' and launched with the slogan 'THIS IS IT. L & M filters are just what the doctor ordered.'

Filtered brands became a must for every manufacturer. The mighty R. J. Reynolds, after fighting a rearguard action with Camel adverts that featured doctors, started work on its own offering. RJR had a philosophical problem with filtered cigarettes. Camel's success had been based on their flavour, and flavour had been a dominant influence in corporate thinking. Filtered cigarettes, however, were comparatively flavourless. It seemed that the public were prepared to sacrifice taste for the perception of safety. RJR gambled against the trend towards tasteless cigarettes, opting to load their new filter tip with tar and nicotine, so that even after the filter had done its work some taste and tar would remain intact. In deference, however, to the modern smoker's less discriminating palates, they decided that quality of taste could be compromised, and that their new filter brand might make use of the 30 per cent or so of tobacco wasted in processing in the manufacture of normal cigarettes. After rigorous experiments with a coffee grinder and a pulp press, RJR came up with RST – Reconstituted Sheet Tobacco – which used all the stems, leaf ribs, tobacco scraps and dust which had hitherto been thrown away. 'Winston', their new filtered wonder, made extensive use of RST. It was offered to the public with the promise 'Winston brings flavor back into cigarette smoking!' and the guarantee 'Winston tastes good, like a cigarette should.'

The introduction of RST marks a change in the cigarette manufacturers' perception of their customers. Cigarettes, despite their origin as poor man's tools, had nevertheless been a genuine tobacco habit. The paper skin that rendered their contents invisible was accepted by both manufacturer and consumer to be at most a necessary evil, but never a cloak of darkness beneath which secrets were concealed. Once manufacturers started treating their products as a package instead of a tobacco delivery system, and a package that had to look prettier or promise better health, wealth or appearance than their competitors' brands, they effectively abandoned the integrity of their product in favour of its appearance.

This happened to be a lucky move. A new, near ideal medium for cigarette advertising had arrived that emphasized appearance alone. Television gave everyone who could afford its receivers a personal cinema in their front room. Although television did not have the formality, did not have

gesture of class solidarity. Hand-made cigarettes were for the well-off, Woodbines, Weights and Tenners for the working class. Only Players, the one with the sailor on the packet, seemed able to cut across the class divide. But branding was not a foolproof division – one could not always see what a man was smoking. Hence the way in which cigarettes were handled became indicative of class. The working man held his fag with his thumb and forefinger with the burning end pointing towards his palm. The middle-class smoker held his cigarette between his first and index fingers with the fire pointing outwards. This stance was indicative of confidence, of a smoker content with their place in the world, and contrasts with the uncertain smoking posture adopted by the workers, who suffered restrictions on where they could smoke in some heavy industries, and for whom smoking was an activity that sometimes had to be concealed.

Mass Observation also noted the aggressive male language employed by working-class smokers towards their habit. 'Their actions, moreover, even more than their language, are frequently clothed in aggressiveness. Some speak of "grinding", "crushing", even "killing" a stub and a favourite trick is to burn it to death in the fire or to drown it in the nearest available liquid.' The aggression manifested itself in a new movement of British literature, which portrayed the young of the age as possessed by an unreasoning rage. The movement's flagship was John Osborne's play *Look Back in Anger* (1956), whose hero Jimmy Porter rails against the purposeless of modern urban existence, while baiting his women and smoking. Here is Jimmy in full spleen, criticizing his elders yet simultaneously envying their security:

The old Edwardian brigade do make their brief little world look pretty tempting. All homemade cakes and croquet, bright ideas, bright uniforms. Always the same picture: high summer, the long days in the sun, slim volumes of verse, crisp linen, the smell of starch. What a romantic picture. Phoney too, of course. It must have rained sometimes. Still, even I regret it somehow,

phoney or not. If you've no world of your own, it's
rather pleasant to regret the passing of someone else's.

The play utilizes a reversal of Victorian class divisions among
smoking implements to assist characterization. Jimmy
smokes a pipe while he rages against and yearns for the
security of a bygone era. His wife and his workmates smoke
cigarettes, and do not seem to care about nor indeed to have
noticed, that civilization is collapsing around them. The frustration embodied by the character of Jimmy Porter was also
depicted on celluloid. In *Saturday Night and Sunday
Morning* (1960) Albert Finney plays Arthur Seaton, a hard
smoking, hard drinking factory worker, who sows his seed
with vigour and whose life is an act of defiance of the married
ideal until he too is trapped by nappies and rent books.

It was not only the working class that felt frustrated by
the apparent pointlessness of their existence. The loss of the
Empire resulted in a similar sense of purposelessness amongst
the middle class. The middle class had to pay to visit India
or Burma after these countries had celebrated their independence ceremonies, and brass bands no longer met their
boats and played them ashore. The Cold War added to the
claustrophobia. Johnny Chinaman and Johnny Russian had
half of the world in their clutches, and were encouraging
chaps to chuck bombs at 'imperialist lackeys and running-
dogs' in other countries. There were few places left to govern,
and many others unsafe to visit, and so the middle class
needed a proxy – a man who could move about the world
as easily and as elegantly as their forefathers had done,
stimulating the envy and respect of foreigners everywhere
they travelled.

Fortunately, their fantasies did not have to go the same
way as the Empire. Fiction provided salvation by inventing
a middle-class version of the angry young man – the
dangerous young man. James Bond, an employee of the
British Secret Service, who refer to him as '007', is a very
Victorian type of hero. Bond works in an urban environment, which he finds stifling – a labyrinth of mediocrity,

where man's potential is suffocated by bureaucracy and petty custom. Like Sherlock Holmes, 007 lives for adventures, but unlike his Victorian predecessor, he is a civil servant, and instead of seeking out excitement, must wait until his employers send him on a mission.

James Bond was the first great post-war smoking hero – a superhuman smoker in an age of smoking giants. He would nonchalantly consume up to sixty cigarettes in a busy evening's gambling in the casinos of Europe. His performance in his debut in print – *Casino Royale* (1953) – was applauded. 'The man every man wants to be and every woman wants between her sheets,' the *Sunday Times* enthused. Although James Bond's creator, Ian Fleming, was always precise as to his characters' appearances, sometimes going so far as to specify their underwear, the attention he pays to their tobacco habits approaches the obsessive. It is a measure of the depth of penetration branding had achieved in the 1950s, and of the information a brand was considered to convey about its owner. Fleming describes Bond's own smoking apparatus meticulously. Bond preferred 'Morland Specials' made up to his own recipe of a blend of Virginian and Balkan weed by a bespoke tobacconist in St James's, although he selects Chesterfields when working in tandem with the CIA. Other smokers in the Bond series follow the usual literary conventions. M, for instance, a wise old sea-dog who is Bond's boss, smokes a pipe. Interestingly, Bond's enemies seldom smoke. Both Dr No and Ernest Blofeld are tabagophobes. Sir Hugo Drax, the triumphantly vulgar villain of *Moonraker*, smokes a common brand of cigarettes and cheats at cards by looking at their reflection on his cigarette case. Bond drubs him, of course, on a mixture of 'Moorland Specials', Benzedrine and champagne. Another threat to global security, Scaramanga, the man with the golden gun, who creates his eponymous weapon from his smoking apparatus, also suffers defeat at Bond's hands, the temptation to play with his smoking equipment indicating a fundamental character defect that enables Bond to overcome him.

The Bond books introduced a new symbolism to tobacco's written heritage – the idea that smoking was dangerous to the smoker. The concept was not a cultural novelty – shamans since time immemorial had occasionally smoked or tobacco-licked themselves to death, and in the days of the Elizabethan reeking gallants, a man's virility was reckoned to be judged by the amount that he could smoke – but this was the first time excessive smoking was presented as a risky business and something only heroes should attempt. Bond's excessive smoking habit is a self-inflicted torture, which enhances his fictional character in the reader's eyes. If Bond is devil-may-care enough to hurt himself, he certainly will be less than kind to his enemies.

The concept reflected in the Bond series that smoking might be injurious to health was gathering advocates outside the pages of fiction. Rather than vanishing like the temperance movement and the Nazis, the issue had grown. Even the most fervent supporters of tobacco were forced to acknowledge its existence. The valedictory pages of Compton Mackenzie's *Sublime Tobacco*, which as its title suggests is an unqualified celebration of La Diva Nicotina, give a nod in the direction of the issue of health: 'I ask myself how many men or women I have seen in all my life of whom I could say that tobacco had been their ruin, and I can confidently answer "none". I could not give the same reply to such a question about alcohol.' Mackenzie, in real life an M figure who had fought in one world war and worked in the secret service in both, offers his own experience as further evidence of tobacco's harmlessness: 'in the course of my life I have smoked 200,000 pipefuls of tobacco at the very least, and probably nearer quarter of a million. The volume of smoke from those pipes might not disgrace Vesuvius when not in full eruption. My memory is crystal clear. My power of concentration is undiminished. My digestion is perfect. My heart is sound.' Fifteen years after completing *Sublime Tobacco*, Mackenzie died in his bed at the age of eighty-nine.

Such Olympian assessments of the benefits of smoking

were seldom made after the 1950s. In *Thunderball*, the ninth book in the James Bond series, 007 is packed off to a health farm, partly on account of his tobacco habit. 'This officer . . . remains basically physically sound,' commences Bond's medical report. 'Unfortunately his mode of life is not such as is likely to allow him to remain in this happy state. Despite many previous warnings, he admits to smoking sixty cigarettes a day. These are of a Balkan mixture with a higher nicotine content than the cheaper varieties. When not engaged upon strenuous duty, the officer's daily average consumption is in the region of half a bottle of spirits.' The results of such excesses, brought on by boredom, the same devil that drove Sherlock Holmes to opium and cocaine, include a furry tongue, blood-shot eyes and a hard liver. Bond is prescribed a fortnight of rest and abstinence, a healthy diet of nut cutlets and carrot juice, which regime, it is implied, is all that is needed to restore him to perfect health. Had Bond's creator intended his adventures to be gritty social commentaries, he would have sent 007 not to a health farm, but to a cancer ward.

16

Tumours

Lung cancer and smoking – of mice and men – health warnings on cigarette packs – tobacco finds an ally in cannabis – tobacco in Vietnam and in China's Cultural Revolution

Cancer wards were still new in 1961, the year Ian Fleming wrote *Thunderball*. They were, however, necessary: deaths from lung cancer had been rising precipitously, in the USA, for example, from 2357 cases in 1933, to 7121 cases in 1940 and to 29,000 deaths in 1956, by which time the lung cancer death rate was 31 per 100,000 – more than nine times the 1933 figure. If a time lag of twenty years was allowed for, the growth in lung cancer matched the increase in cigarette smoking. The scientific community found this resemblance startling, and had been researching whether the two statistics might be related from the 1940s onwards. In 1950, landmark studies on both sides of the Atlantic concluded that the concurrent growth of cigarette smoking and lung cancer was no coincidence. In the 27 May 1950 issue of the *Journal of the American Medical Association (JAMA)*, an epidemiologist named Morton Levin published the first American study confirming a statistical link between smoking and lung cancer. After surveying the smoking habits of 236 cancer hospital patients, he found the risk of lung cancer to be ten times as high for heavy, long-term smokers as for non-smokers. In the same issue of *JAMA*, another study entitled 'Tobacco Smoking as a Possible Etiologic Factor in Bronchiogenic Carcinoma: A Study of 684 Proved Cases', by Ernst Wynder and Evarts Graham, found that 96.5 per cent of lung cancer patients interviewed were heavy-to-chain-smokers. In September of the same year two Britons, Richard Doll and A. Bradford Hill, published a report entitled 'Smoking and Carcinoma of the Lung' in the *British Medical Journal*. After studying 1732 cancer patients and comparing them with 743 non-cancer patients, they arrived at the statistic that heavy smokers were fifty times as likely as non-smokers to contract lung cancer: 'smoking is a factor, and an important factor, in the production of carcinoma of the lung,' they concluded.

This clearly was important news for smokers and tobacco manufacturers, not to mention the governments who regulated tobacco products and, in many European countries such as France and Spain, ran tobacco monopolies. However, a further seven years were to pass before any official action occurred in any country, during which period of time a large number of further statistical analyses were carried out – all of which arrived at the same conclusion – that smokers were significantly more likely to die of lung cancer than non-smokers. In 1957, the US Surgeon General Leroy Burney issued a *Joint Report of Study Group on Smoking and Health*, which stated that, 'prolonged cigarette smoking was a causative factor in the etiology of lung cancer'. This was the first time the American Public Health Service had taken a position on the subject. The British went further. In the same year, the British Medical Research Council published *Tobacco Smoking and Cancer of Lung*, which declared that 'a major part of the increase [in lung cancer] is associated with tobacco smoking, particularly in the form of cigarettes' and that 'the relationship is one of direct cause and effect'. Meanwhile, the incidence of lung cancer continued to climb. In 1957, it claimed the lives of Humphrey Bogart, perhaps the most memorable screen smoker of all time, and of Dr Evarts Graham, the chain-smoking co-author of the landmark 1950 study connecting smoking and the disease.

Reports linking tobacco and cancer did not come as a surprise to tobacco manufacturers, particularly in the United States, where they had been shadowing health issues with product developments for over a decade. The manufacturers had also undertaken various PR manoeuvres intended to bolster public confidence in smoking, one of which involved them holding hands in public. In 1954, the US tobacco companies, acting in harmony for the first time since the dismemberment of Buck Duke's monopoly, issued a joint pronouncement on the health aspects of smoking to their valued customers. Their 'Frank Statement to Cigarette Smokers' was placed as an advertisement in 448 newspapers across America, reaching a circulation of 43,245,000 people in 258 cities. It aimed to portray the tobacco companies as responsible corporate citizens who were shocked at allegations that their products might cause cancer and commenced with righteous denial:

RECENT REPORTS . . . have given wide publicity to a theory that cigarette smoking is in some way linked with lung cancer in human beings.

Although conducted by doctors of professional standing, these experiments are not regarded as conclusive in the field of cancer research. However, we do not believe the results are inconclusive, should be disregarded or lightly dismissed. At the same time, we feel it is in the public interest to call attention to the fact that eminent doctors and research scientists have publicly questioned the claimed significance of these experiments.

Distinguished authorities point out:

That medical research of recent years indicates many possible causes of lung cancer.

That there is no agreement among the authorities regarding what the cause is.

That there is no proof that cigarette smoking is one of the causes.

That statistics purporting to link cigarette smoking with the disease could apply with equal force to any one of many other aspects of modern life. Indeed the validity of the statistics themselves is questioned by numerous scientists.

We accept an interest in people's health as a basic responsibility, paramount to every other consideration in our business.

We believe the products we make are not injurious to health.

We always have and always will cooperate closely with those whose task it is to safeguard the public health.

The 'Frank Statement' concluded by announcing that the manufacturers would establish the 'Tobacco Industry Research Committee', run by 'a scientist of unimpeachable integrity and national repute', which would aid and assist 'the research effort into all phases of tobacco use and health'. The Tobacco Industry Research Committee's first scientific director interpreted its aims more widely as 'research and public relations. Our job is to maintain a balance between the two.' The committee appears to have focused on the good news, i.e. public relations. Although the manufacturers were aware of the growing case against their product, in contrast to their united disclaimer, their in-house reactions to cancer were less definite. For instance, in 1957 British American Tobacco could not even bring itself to name cancer in internal memos, referring to it instead under the code name 'zephyr': 'As a result of several statistical surveys, the idea has arisen that there is a causal relation between zephyr and tobacco smoking, particularly cigarette smoking.' Three years later, the American manufacturer Liggett & Myers had progressed to the stage of privately acknowledging that their product had some undesirable side effects, as a 'confidential' memorandum from a consulting research firm hired to probe the mysteries of smoking revealed: 'There are biologically active materials present in cigarette tobacco. These are: (a) cancer causing, (b) cancer promoting, (c) poisonous,

(d) stimulating, pleasurable, and flavorful.' Such lapses aside, however, the tobacco manufacturers responded to assertions that their product might be lethal with denials, by discrediting their opponents, and with litigation.

American tobacco manufacturers were not only under direct attack from scientists, but also their customers, who resented being given lung cancer by their favourite consumer product. An early law suit, Green vs. American Tobacco Co., brought by an unlucky Lucky Strike smoker in 1960, seemed to point the way towards a tobacco company's liability to its customers. The plaintiff died of cancer mid-case, but the lawsuit was continued by his lawyers, and at its conclusion Miami Federal District Judge Emett Choate asked the jury to consider: (1) was cancer primary in the lung? (2) Did this cause his death? (3) Did the smoking of Lucky Strikes cause his cancer? In all three instances, the jury voted 'yes'. The fourth interrogatory asked, 'Did the cigarette company have knowledge of the harmfulness?' The jury said, 'no'. A retrial was held and American Tobacco exonerated on the basis that its products simply were not dangerous enough. Other lawsuits against other tobacco companies surfaced, but most foundered on causation, i.e. the difficulty of proving that cigarettes were responsible for the cancer in question. Sure, the plaintiff was dying of a tumour; sure, that other smokers had died of them, but that was irrelevant, like bringing up the *Titanic* as evidence every time someone drowned in their bath. The tobacco companies quickly became fearsome courtroom opponents, using every procedural trick available to wear down plaintiffs, many of whom gave up their cases because they died or ran out of funds.

Although lung cancer was still rare, by the time of Green vs. American Tobacco it had a public reputation as a killer. Cancer's deadly potential, the extreme suffering its victims underwent before death, and above all its incurability, stigmatized the disease, and its victims. Despite the advances in medicine and surgery, cancer remained an official mystery, and a private hell.

Description of the actual processes of lung cancer resists

simplification. Its causes were not known, but were believed to result from irritation of the lungs. The irritation caused cells in the lining of the lungs to divide, in order to create a defensive coat against the irritation, in a process called hyperplasia. But sometimes this defence mechanism over-reacted and gave rise to mutations. These mutant cells, unlike normal cells, had no limit to the number of times they would divide. Their presence marked the second stage of the cancer cycle, known as metaplasia. In the third stage, called neoplasia, the mutant cells clumped together, effectively colonizing the lung, drawing nourishment from the blood and utilizing the sufferer's lymphatic system, at the expense of normal cells and normal body function. The cancerous cells could also colonize different parts of the body, including the brain, the bones and the lymphatic system. There are over a hundred forms of cancerous cell, not all of which are malignant, i.e. flourish at the expense of normal body function. Lung cancer is generally grouped into two major categories, based on the physical appearance of the cancer cell under the microscope. The two types are 'small-cell lung cancer', and 'non-small-cell lung cancer'. Non-small-cell lung cancer comprises about 75 per cent of all lung cancers. It generally grows at a moderate rate, and usually spreads to the adjacent lymph nodes before travelling to other parts of the body. There are three basic subclasses of non-small-cell cancers: squamous cell, large cell and adenocarcinoma. Squamous cell is the type most frequently seen in smokers and is the most likely to metastasize, i.e. turn into killer cells. Adenocarcinoma is the lung cancer most frequently found in non-smokers, but occurs in smokers as well. Large cell carcinoma is the one most likely to spread to the brain.

The effects of lung cancer on the sufferer are horrible. The tumours cause excruciating pain wherever they are colonizing. As healthy cells are starved to feed cancerous ones, the victim becomes haggard and wasted, and as less healthy lung is available for breathing, they are suffocated at the same time as being tortured to death.

In order to establish a casual link between smoking and

lung cancer, scientists had to prove that something in cigarette smoke irritated human lungs to the degree that they
started the cancer cycle. Initial investigations focused on
nicotine, the chemical 'soul' of tobacco. Research into nicotine had advanced since the days of killing cats and dogs
with overdoses. The compound's effect on the human body
had been monitored and quantified – nicotine was a heart
stimulant, it mimicked and stimulated chemical transmitters
in the brain, it was addictive, in that people liked its effects
and usually wanted more, and it was still poisonous in large
doses. But did it irritate?

Early research with nicotine yielded disappointing results,
and the search for causation switched its attention from
nicotine to the smoke itself. Cancer had grown not with
smoking, but cigarette smoking, i.e. inhalation. Immediately,
the potential of establishing a link between smoking and
lung cancer became more promising. It was relatively easy
to prove that smoking irritated the lungs. It certainly irritated the eyeballs, which are made of tougher stuff than
lungs. The next step was to prove that this irritation caused
cancer. This final link in the chain was a challenge to forge.
An understanding of the causation of cancer was beyond
the limits of contemporary medicine, although these were
subsequently expanded when asbestos workers started dying
in numbers. The wonder substance in Kent's 'micronite' filter
turned out to have had more sinister properties than the
average fire retardant. A majority of workers at the factory
which had made the asbestos insert for the filters ultimately
succumbed to lung cancer, and their relatives were later the
first to win money from a cigarette company for causing
lung cancer, albeit not with tobacco.

A breakthrough in causation seemed to come in 1953,
when research carried out with mice appeared to prove the
link between cancer and smoking. The mice were shaved and
their backs painted with the extract of Lucky Strike cigarettes. Some of the mice developed tumours, but the results
turned out to be inconclusive upon further examination. The
test mice were tumour prone, whether their skin had been

painted or not. And the concentration of tars and nicotine in the paint was so high that it cast further doubts on the results. The experiment was repeated several times in the ensuing years. Thousands of thoroughbred mice in laboratories all over the United States were shaved and painted, without anyone succeeding in duplicating the 1953 result.

Despite the overwhelming, statistical evidence against cigarettes, and advances in the area of causation, it was not until five years after the American surgeon general's and British Medical Research Council's reports that any government committed itself to advising its citizens, in an unambiguous manner, that smoking cigarettes was a bona fide health hazard. On 7 March 1962, the British Royal College of Physicians of London published a report entitled *Smoking and Health*, intended to prove 'the overwhelming case against tobacco'. It sold out immediately and a further 20,000 copies were sold within the next six weeks. *Smoking and Health* was front page news in the British press. The issue of official intervention was debated in the editorials of the *Guardian* and *The Times*, advice was given on how to give up smoking, and there was speculation as to the ultimate body count of the cigarette peril. Pipe and cigar smokers pointed out their prior warnings in the letters' pages.

The American equivalent appeared two years later in the form of the 1964 surgeon general's report, *Smoking and Health: Report of the Advisory Committee to the Surgeon General of the Public Health Service*. The document's revelation to the public was stage managed in a way which reflected the shock value the authorities considered it to hold. The date chosen for the surgeon general's report's release was a Saturday morning to guard against fall-out on Wall Street. The first two copies of the 387-page, brown-covered report were hand delivered to the West Wing of the White House at 7.30. At 9 a.m., accredited press representatives were admitted to an auditorium in the State Department and 'locked-in', without access to telephones. Surgeon General Luther Terry and his Advisory Committee took their seats on the platform. The report was distributed and reporters were allowed ninety minutes to read it. Questions were answered by Dr Terry and his committee members. Finally, the doors were opened and the news was spread. For several days, the report furnished newspaper headlines across America and led television newscasts. It was ranked among the top news stories of 1964.

American tobacco manufacturers responded with fury to the accusations of death-dealing inherent in the epidemiologists' and surgeon general's reports. The tobacco companies considered themselves an essential part of America's heritage. Americans spoke English because of tobacco. Americans had smoked before they learned to bake apple pie. The country's oldest communities had been founded to grow the weed, and its independence had been bought with it. Insulting tobacco was as unpatriotic as spitting on the Stars and Stripes, and it made a bad impression on the union's two newest members, Alaska and Hawaii, which become officially American in 1959.

In addition to displays of indignation and denials, the cigarette industry counter-attacked. It was used to health scares – it had seen off Thomas Edison, Miss Gaston and the nation's confectionery manufacturers, and it brought the experience gained in earlier battles to confront the voice of American medicine. In a typically forthright manner, an industry sponsored writer suggested tobacco control advocates should obtain psychiatric certification that they were not suffering from pyrophobia or suppressed fear of the 'big fire', i.e. atom bomb. In addition to ridiculing their adversaries, they also set out to discredit them, and here they had a point. Almost all the studies linking smoking and lung cancer were statistical. Despite those inconclusive mice in 1953, no one had yet shown how smoking caused cancer. The manufacturers relied on this defect. Without a causative link, statistics were just numbers. So what if smokers died of cancer? The wheel of karma did not always spin evenly. Next time round it would be the non-smokers' turn. The quality of their opponents' research was scrutinized and any weaknesses exposed. When a later horror report appeared linking smoking to birth defects and claiming it had an adverse effect on the size of babies, Joseph Cullman of Philip Morris remarked that 'some women would prefer having smaller babies'. The mouse painting programme was also attacked. A tobacco industry spokesman pointed out that tomato ketchup would have caused tumours if mice were coated with it in the right concentrations. Interestingly, tobacco manufacturers were used to their opponents slaying mice. Henry Ford's *Case against the Little White Slaver* included a testimonial from a Mrs T. E. Patterson, president of the Georgia Woman's Christian Temperance Union, in which she had admitted to killing 'dozens of mice' with cigarette products.

However, as the law suits from unhappy dying smokers increased, tobacco company representatives became more guarded in their approach to the cancer issue. When two Englishmen, Sir Philip Rogers and Geoffrey Todd, embarked on a fact finding mission to the USA in 1964, they reported that 'the leadership in the US smoking and health situation . . . lies with the powerful policy committee of senior lawyers advising the industry, and their policy, very understandably . . . is "don't take any chances".' The only controversial comment Sir Philip and Mr Todd heard from an industry figure on the health issue was that 'obviously there were some people who should not smoke, e.g. those with emphysema'. Instead of talking medicine, the industry had switched its focus to fun and flavour. 'Camel time is pleasure time for you' as R. J. Reynolds expressed it.

Cigarette manufacturers concentrated their counter-attack on promotion. They hiked their advertising spend and pushed cigarettes in every way they and their highly paid advisers could imagine that did not involve cancer. Philip Morris led the way with its revamped Marlboro filter brand. The brand's totem Marlboro Man, like Frankenstein, had come to life. Propelled to fame by an immense advertising budget, Marlboro Man began to ride nightly across American TV screens in 1964. He lived in a giant, scenic landscape named 'Marlboro Country', and viewers' hearts leaped as the beauties of their own fair land were revealed. Marlboro Man did not look like he would succumb to lung cancer. The associations were all wrong. Marlboro Man breathed clean air. His diet and lifestyle were simple and healthy. He was never shown whipping slaves or reliving equivocal parts of America's heritage. Who would you rather trust? Someone you had never heard of in a white coat who spent their life wandering around hospital wards counting dying people, or Marlboro Man?

Quick on the draw

The US government soon realized that it was not enough to educate people as to the risks inherent in smoking. Their campaign was simply too drab to compete with the colour and excitement of tobacco advertising. It was hard for the average American to believe such a heavily promoted product as cigarettes might be fatal. The citizens were behaving like dazed children – they lost their heads the instant they were confronted with the glamorous aspects of smoking. They thought of fun instead of lung cancer. The only way to shake them out of this trance was to put warnings on cigarette packets. By juxtaposing health and pleasure people would be forced to consider both issues each time they purchased.

The concept of health warnings on cigarette packets had been in the air in America since 1957, when a bill had been introduced to the Senate requiring all cigarette packs to carry the label: 'Warning: Prolonged use of this product may result in cancer, in lung, heart and circulatory ailments, and in other diseases.' The bill was killed by tobacco industry supporters, presumably on the grounds of aesthetics. Cigarette packets were masterpieces of design and should not be defaced by crude statements of fact. By the mid sixties the government had worked up the courage to insist its citizens should be reminded of the risks every time they smoked. The tobacco companies stonewalled, hankering back to the days when you listed health benefits on a pack of cigarettes. The resulting federal Cigarette Labeling and Advertising Act, which came into effect on 1 January 1966, was a fudge – a government warning in the tobacco companies' words: 'Cigarette smoking may be hazardous to your health.' So what? So might everything. People drowned in their baths, but would probably still take them if they carried a similarly equivocal warning. The American media noticed the compromise. The *New York Times* described the cigarette labelling act as 'a shocking piece of special interest legislation – a bill to protect the economic health of the tobacco industry by freeing it of proper legislation'. The *Atlantic Monthly* reflected that the cigarette industry had 'found the

best filter yet – *Congress'*. The message had been eviscer-
ated, but the protection it offered the tobacco companies
was solid. If a customer got cancer it was their own fault.
They had, after all, been warned.

US cigarette sales responded positively to warning labels.
They rose by over 16 billion in 1965, and by a further 7.8
billion in 1966. Ideally, everyone should have been happy.
The government had done its duty towards its citizens by
giving them warning labels and the education to read them,
the manufacturers had done their fair share by reminding
their customers that although smoking was fun and flavour-
some, it might carry health risks, and the customers had
made informed choices in buying more cigarettes. The advent
of health warnings appeared to take some of the heat out
of the debate as to the merits of cigarettes. In the same year
as they appeared, the US government still had enough confidence in the little white slaver to send 600 million cigarettes
as a gift to flood disaster victims in India.

The British government adopted a different approach to the
Americans towards the duty of advising its subjects of the
dangers of cigarette smoking. Official thinking seems to have
been that the average Briton was dead to reason once their
sense of pleasure had been provoked. By the time Her
Majesty's subjects had cigarette packs in their hands it would
be too late for any warning. Better, therefore, to deny them
temptation by restricting advertising. In 1962, following the
Royal College of Physicians' *Smoking and Health* report,
the postmaster general took measures against TV advertising, banning the association of happiness, virility, love
and/or adventure with smoking. In 1965, all tobacco advertising was banned on television (coincidentally, a useful blow
against private competition to the BBC). British cigarette
sales responded positively, rising to 112 billion that year,
and to 118 billion in 1966. The adverts had gone, but people
still smoked on television. Tobacco had friends in high places.
There were many living witnesses to its cerebral powers
appearing nightly on Great Britain's TV screens. These

included the Prime Minister, Harold Wilson, who had promised to improve the country with the 'white heat' of technology, by which he meant a washing machine bought on hire purchase in every household, and who was elected 'Pipesmoker of the Year' for 1966.

The task of getting the anti-smoking message across was by and large left to Britain's local authorities. They were given money and a million government printed posters and told to get on with it. They ran into immediate problems with the posters from the poster advertising industry. Tobacco companies were a mainstay of poster advertising, and it was more than the poster companies' jobs were worth to offend them. Nevertheless, the local authorities soldiered on with workshops, pamphlets and ill-attended meetings. Central government occasionally lent some aid. Two mobile anti-smoking units were launched by the health minister, Enoch Powell, in 1962, which during the next two years travelled over 125,000 miles of British roads. In general, however, these well-intentioned efforts suffered from poor presentation. The councils laboured away at the populace with statistics, instead of images, and made little impression on British smokers.

Meanwhile, the UK tobacco industry was fighting back, pioneering new ways of promoting cigarettes. These included subsidizing the sport of motor racing, which had come a long way since the It Girl Agatha Runcible's adventures with car no. 13 in *Vile Bodies*. In 1968, Colin Chapman's Team Lotus became the first Formula One outfit to accept tobacco sponsorship. Motor racing and smoking seemed made for each other. Both were glamorous, dangerous and fun. The cars looked great in tobacco company liveries and the race tracks constituted giant billboards for brand promotion. Despite Henry Ford's campaign against them, cigarettes had grown up with the motor car – ashtrays were standard dashboard fittings for Jaguar E-types and Minis alike – and the tobacco companies exploited this association. Cigarettes, like motor cars, were symbolic of independence and success.

The real problem with the anti-smoking crusade was that Britons were having a good time in the 1960s, enjoying and refining the pleasures of affluence young Americans had discovered the decade before. Something of the blitz spirit lived on and the prospect of being killed by lung cancer in greater numbers than Nazi bombing raids had achieved was measured and ignored. Britons cared less for their state of health than Americans, believing it to be a necessary evil. Furthermore, the release of statistical proof that smokers were more likely to die of lung cancer than non-smokers coincided with explosive growth in the youth market, to whose icons the proven danger of smoking was an attraction, and the nascent, youth-orientated anti-smoking programme something else to rebel against. Fashion was ranged squarely against science – Doll and Hill vs the Beatles, and the absence of non-smoking idols told heavily against the men in white.

The Beatles smoked, initially, only cigarettes, and did much to improve the habit's reputation in the eyes of their fans. When they were joined on the world stage by other British bands including the Rolling Stones, Cream and the Small Faces, all of whom sought to appeal to the rebellious side of youth, tobacco smoking received a fresh endorsement. It also enjoyed support from a new form of smoking rapidly gaining popularity in the industrialized world.

Throughout tobacco's history in the West it had been the king of smokes. Few people had considered putting any other weed in their pipes. When another combustible plant began to enchant the people in numbers, the world of smoking was revitalized with an injection of transcendental associations. Although 'pot' (*Cannabis sativa*) had been an inspiration to Western culture since Neolithic times – some of the spiral maze rock carvings of Gavrinis in Brittany (*c.* 4000 BC) have been attributed to its influence – its power to stimulate was seldom acknowledged, let alone celebrated. The use of cannabis spread dramatically in the 1950s and 1960s, and it gave rise to an alternative smoking habit, with its

own subculture and icons (including a talking cat) that
supplemented those which had grown up around tobacco
smoking since the days of Columbus. Interestingly, cannabis
'joints' were usually shared, and the informal ceremonies
that developed round their communal consumption are reminiscent of Amerindian societies, where smoking was the principal leisure activity, and whose members would share a pipe
before discussing the pleasures of a forthcoming beaver hunt,
or what the stars were for.

When the market for cigarette rolling papers started
growing exponentially, British tobacco companies took an
excited interest in this phenomenon. Why did so many young
people want to roll their own cigarettes? What sort of
tobacco were they using? They decided they were witnessing
a rebirth of the Sevillian *papelote* and began to sell cigarette
tobacco loose, and to pay attention to the branding,
marketing and packaging of hand-rolling tobacco. Adverts
for this class of product celebrated the ingenuity of its
customers in making their own cigarettes. Images of
craftsmen and artisans were employed so that the hand roller
might associate wrapping tobacco in paper with a higher
plane of creativity.

American manufacturers were quicker to identify what
it was all the young people were smoking and contemplated
opening up another product line. Cannabis cigarettes would
take some of the pressure off tobacco. Scientists would have
to start painting mice all over again. Sadly, the plans for
cannabis cigarettes had to be shelved when the substance
was made illegal. As it happened, scientists had progressed
from mice in their quest for the link between smoking and
cancer. They were teaching dogs to smoke. In May 1967,
Oscar Auerbach had installed ninety-four pedigree beagles
in a laboratory, for the purposes of establishing whether he
could give them cancer by making them smoke. The beagles
had had holes cut in their throats to allow special smoking
collars to be fitted. After an 875 day period, in which some
dogs smoked filtered cigarettes, others unfiltered and the

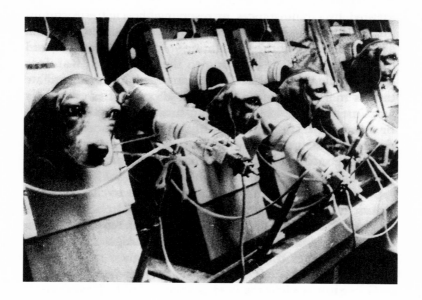

This is not a pipe

control dogs nothing, all the surviving dogs were killed and their bodies cut open to check for tumours. The results were promising, but just not enough of the smoking dogs had got cancer to prove conclusively that smoking cigarettes caused lung cancer in humans. As a tobacco company executive pointed out: '[W]e believe that the Auerbach work proves beyond reasonable doubt that fresh whole cigarette smoke is carcinogenic to dog lungs . . . although it does not help us directly with the problem of how to modify our cigarettes.'

Never before in the cigarette's history had so many powerful and antagonistic forces been at work: Beatles, beagles, health warnings, flavour, pain and glamour were all spinning round the average smoker's head. The popularity of cannabis further confused the issue. Was that bad

17

Blowing Bubbles

*Triumphs and tribulations for pipe and cigar
smokers – smokers in space and on the
moon – a decade of pain in Great Britain –
cigarette manufacturers target children –
nicotine and addiction – tobacco enriches its
opponents, and consoles its victims*

Most research into smoking and disease had focused on cigarettes, leaving the older forms of tobacco consumption, pipe and cigar smoking, relatively free of criticism, if not entirely unscathed. The US surgeon general had stated that whereas heavy cigarette smokers had a 68 per cent higher mortality rate than non-smokers, the corresponding figures for cigar and for pipe smokers were 22 per cent and 12 per cent respectively, implying a greatly reduced risk for these traditional methods of smoking. The news was even better with regard to lung cancer: cigar smokers were only twice as likely to die from it than non-smokers, compared to the twenty-five fold fold risk run by chain-smokers of cigarettes.

While pipe and cigar smokers did not have to suffer the indignity of having health warnings engraved on their pipes or printed on to cigar bands, they were nevertheless subjected to the same restrictions that were beginning to affect their cigarette smoking brethren. Although the No Smoking signs going up everywhere showed a red circle enclosing a cigarette with a red line through it, they did not imply that other forms of smoking were permitted. Indeed, pipes and cigars were banned on many forms of public transport long before cigarettes were prohibited. The bans were usually on the grounds of smell as opposed to health. While the human nose was unchanged, the category of odours considered offensive had expanded. The industrialized world had developed a fascination for personal hygiene. Although it was acceptable to spew vast quantities of complex and toxic chemicals into the environment, dirty teeth or smelly armpits were believed to be barbaric, and in a society increasingly obsessed with appearance, the useful bacteria which had hitherto flourished on mankind's skin and various soft tissues were banished to the history books. The principles laid down by Beau Brummell and his fellow dandies were adopted by the common man and woman, and their communal fixation with smelling fresh and looking clean was catered to by tooth-paste, deodorant and shampoo manufacturers, who, eager to push forward the frontiers of their market, introduced products people never even realized they had needed including hair conditioner and vaginal deodorants. Some products turned out to be nearly as dangerous as the cigarettes whose smell they were perhaps intended to suppress.

While pipe and cigar smokers suffered from being clumped together with cigarette smokers, they enjoyed a few of their own tragedies and triumphs in the 1960s and 1970s. Cigar smokers were provided with their own cowboy icon, when Sergio Leone recast Marlboro Man as a murderer and a cigar smoker in a series of films that chronicled the adventures of Clint Eastwood as 'the man with no name'. In films such as *The Good, the Bad and the Ugly* and *A Fistful of Dollars*, the straight shooting, straight spitting cigar smoker of old was resurrected, and so appealing was this vision that demand for cigars in the United States increased. Tragedies of the period included a shortage of Havana cigars. Cuba was slipping into the arms of communism, and, after nearly causing an atomic war in 1962, its government, under the guidance of Generalissimo Fidel Castro, declared an official communist party in 1965. Despite the Cubans' reputation for knowing how to enjoy themselves, the new party did not mean fun for all – indeed, if it promoted any pleasure it was the Victorian one of self-denial.

Western cigar smokers had to endure the indignity of their favourite capitalist success token – the Havana – falling into communist hands. Cuban production was centralized in a few large factories, each producing a variety of brands, where cigar rollers were read propaganda while they worked instead of the traditional romances. Some of Cuba's oldest cigar-making dynasties fled the country, taking their brand names with them. The brands that left included Henry Clay, named after the Great Compromiser, the nineteenth-century champion of emancipation and equal rights, who had no place in communist Cuba. In some cases, the brands' owners set up in competition against their own labels that had been appropriated by the state. Partagas, Romeo y Julieta, H. Upman, Punch and Hoyo de Monterrey all went into production in exile in the Dominican Republic and in Honduras, as rivals to Cuban-made cigars of the same names.

Cigar smokers did not suffer the same persecution as cigar manufacturers in post-revolutionary Cuba. Castro and his fellow Argentinian revolutionary, Che Guevara, were fervent smokers, and both believed the habit to be a force for good. Che was inspired by the ideal he had formulated of a 'new man' – a moral and collectivist-minded revolutionary who smoked. As Che pointed out in his practical guide to staging a revolution (*Guerrilla Warfare*, 1961): 'a customary and extremely important comfort in the life of the guerrilla fighter is a smoke'. While Che was shot like a dog in the Bolivian jungle in 1967, Castro lived on to enjoy the delights of a class free society, which had cleansed itself in the flames of revolutionary warfare and had no fear of smoke, or ash. As the Generalissimo himself observed: 'There is nothing in the world more agreeable than having a place where one can throw on the floor as many cigar butts as one pleases without the subconscious fear of a maid who is waiting like a sentinel to place an ashtray where the ashes are going to fall.'

Castro's weakness for cigars, and willingness to take them

from strangers, nearly led to his downfall. The American CIA
had offered a bounty of $150,000 for his assassination, and
a plan was laid to poison Castro via his cigars. Despite a
continuing torrent of tobacco-is-a-killer statistics from their
own government, the CIA were not content to let nature run
its course. They treated a box of Castro's favourite smokes
with botulinus toxin, after this poison's lethal effect had
been tested on monkeys, then entrusted the spiked cigars to
an agent to offer to the Generalissimo. They failed to reach
their target, and whether they ever contributed to the casualty register of dead smokers is not recorded.

Despite its failure to overthrow Castro, the US government still cared about giving every patriot good news. When
the Russians put a satellite (Sputnik) and a dog (Laika) into
space, the USA took up the challenge. They decided to put
not just pets and metal chunks but Americans in space, and
on the moon as well. They succeeded in their first objective
in 1962 when John Glenn, a cigar smoker, became the first
American to orbit the earth. He was presented with his own
weight in cigars upon his return from space. The United
States' second off-world objective was achieved in 1969
when Space Commander Neil Armstrong, climbed down
out of the Apollo 11 mission's lunar module and stepped
on to *luna firma*. He was later joined by Space Captain
Buzz Aldrin on the face of the ancients' Goddess of Love.
Commander Armstrong was a pipe smoker in the true
explorer's mould, although he did not take his instrument
of pleasure with him into space. Nevertheless, this small step
by a smoker, yet giant for mankind, was a measure of the
distance the weed had travelled from humble scrub to accom-
panying humanity (internally) to the moon and back.

Meanwhile, back on Planet Earth, restrictions against
smokers and their suppliers continued to multiply, especially
in the USA. On 2 January 1971, cigarette advertising was
banned on American television. The sudden termination of
Marlboro Man's TV career was followed by further indigni-
ties. In 1972, all cigarette adverts in America were required to
carry health warnings. In addition, non-smokers had become

more vocal and were demanding and getting non-smoking areas which they could call their own, in which smokers were not just unwelcome, but prohibited. Having established the principle of segregation, the non-smokers set out to reverse the presumption that the world belonged to smokers, bar a few non-smoking sanctuaries. They achieved their initial success in the state of Minnesota, whose 1975 Clean Indoor Air Act was enacted to protect 'the public health and comfort and the environment by prohibiting smoking in public places and at public meetings, except in designated smoking areas'. The Clean Indoor Air Act was the first to presume that smoking was forbidden unless specifically authorized.

Moreover, instead of merely advising people to stop smoking, and restricting where they could smoke, official bodies were telling them how to stop. The tobacco industry spawned another industry, as psychoanalysts and snake oil salesmen started earning money from teaching people to give up smoking, or sympathizing with them when they could not. Giving up smoking books appeared on the best-seller lists, whose patented methods for renouncing the weed bore the heritage of the temperance movement and of Dr Freud. The temperance-influenced tomes advised the smoker not to besmirch God's gifts – the body was a temple, not a chimney. The type of giving up smoking book that drew its idiom from psychoanalysis saw the habit as a matter of super-ego control that could be achieved through positive thinking.

In Great Britain, giving up smoking counselling was already providing work for a number of state employees. Ever since the 1957 publication of statistics linking cancer to smoking, the British medical establishment had treated the act of smoking as a disease in itself and sought methods of curing it. The exotic ailment was initially placed under the care of the London School of Hygiene and Tropical Medicine, but when the local authorities took on the duty of advising Britons of the perils of smoking, they too began to experiment with give up smoking courses. By 1970 a number of private operators had joined them in the field. No single course had a success rate as high as 50 per cent. Ex-smokers often resumed their habit, sometimes after years of abstinence: 'It was on a Monday . . . I was at the Phene Arms and there was a girl at the bar smoking cigarette after cigarette, filling up the ashtray . . . I wanted to talk to her so I asked her for one . . . I was back on a pack a day the next morning.' Doctors even recorded examples of heart attack victims resuming smoking before they had left hospital.

The 1970s proved to be an ugly time for Great Britain, its smokers and its cigarette manufacturers. The manufacturers had begun the decade flush with optimism. A British American Tobacco researcher, Colin C. Greig, had summed up their goals by quoting Oscar Wilde: 'A cigarette is the perfect type of a perfect pleasure. It is exquisite and it leaves one unsatisfied. What more can one want?' before observing, 'Let us provide the exquisiteness and hope that they, our consumers, continue to remain unsatisfied. All we would want then is a larger bag to carry the money to the bank.'

The optimism did not last. Nineteen seventy-one was an *annus horribilis* for the British tobacco industry. A gripping sequel to *Smoking and Health* appeared from the Royal College of Physicians, *Smoking and Health Now*, which described smoking as 'this present holocaust'. As a voluntary concession, cigarette manufacturers agreed to put warning labels on their packs. The BBC ran a special edition of *Panorama*, to coincide with the second RCP report, which highlighted its estimated annual death toll of 100,000 British smokers. An anti-smoking organization, Action on Smoking and Health, acronymed ASH, was founded with government money. *Which?* magazine published a survey of tar and nicotine yields for standard brands of cigarettes, which made unpleasant reading for smokers who had switched to filter cigarettes in an attempt to reduce their intake of these poisons. A group of *Daily Mirror* journalists gave up smoking altogether and reported regularly to their readers on progress. An interdepartmental group of government officials launched a special investigation into cigarette smoking and health which concluded that, all things considered, smoking earned the government more money than it cost in terms of disease and disability. The government put up taxes as a gesture to conscience.

After a bad start to the 1970s for the tobacco companies, their market continued to deteriorate. Worst of all was the discovery that the fight seemed to have gone out of their customers. Although they still smoked, their hearts were treacherous. A survey showed that most smokers believed public places, including sporting facilities, should have non-smoking areas set aside, and that many smokers felt guilty about their habit. In 1972, the proportion of the adult male population who smoked fell below 50 per cent for the first time in over half a century. The percentage of women smoking also broke its promising trend upwards and began to slide. Furthermore, after the *Which?* report on tar yields, smokers deserted their traditional brands in droves, and a niche market cigarette named Silk Cut after the favourite tobacco of the Chinese Imperial court, saw its sales quadruple immediately.

The seventies were the 1960s prematurely aged. Things that had been fun in the sixties were being called dangerous the following decade. Half the innovators of rock 'n' roll were in their graves, though none yet of tobacco smoking related diseases. 'Time takes a cigarette,' sang David Bowie in his elegy to the fallen, 'Rock 'n' Roll Suicide'. Those that had not died were judged by their fans to have 'sold out', a process described by Pink Floyd in 'Have a Cigar'. The gloomier music reflected a gloomier populace. In addition to the official news that the country was full of dying smokers who were creating a greater annual body count than the nation had suffered in World War II, British subjects also had to endure energy and labour crises. Economics as much as warnings contributed to a fall in tobacco sales. The public had begun to count the cost of smoking, and Britain did not have a Marlboro Country to sustain its smokers' dreams – most of its wildernesses had vanished with the Enclosure Acts and the rest were being ruined by a plague of motorways. A measure of the misery of the times was demonstrated by the irony employed in Silk Cut adverts in cinemas, which featured a blacked-up Zulu offering besieged redcoats the opportunity to switch to a lighter brand before they were slaughtered.

The final woe of the decade for British manufacturers was the return of the Americans to the British market. As usual, their shock troops consisted of price cuts and propaganda. Hollywood was on their side and when the disco craze swept the UK in 1977 fans rushed to buy the brands smoked in *Saturday Night Fever*. Philip Morris took especial interest in its founder's birthplace. It was a sponsor of motor racing back home and started backing the European variety. Their Marlboro MacLaren team won the 1976 Formula 1 championships, with England's James Hunt in the driver's seat. Hunt was a perfect cigarette company role model, who was often captured smoking on the podium whilst spraying assorted blonde admirers and crowd members with a magnum of champagne.

Fortunately for the home team, the late seventies provided distractions in the form of other monsters, that appeared greater demons in the public eye than cigarette smokers. These included punk rockers who pierced their bodies, injected smack and sniffed glue, all this in a jubilee year. Happily, punk rockers smoked cheap British brands. The incomparable Sid Vicious of the Sex Pistols paid more attention to his cigarettes than to his bass guitar on stage, which did much for their popularity among the punk community. Smoking also continued its centuries-old association with literary Britain. As the playwright Dennis Potter observed: 'Nobody has yet been able to demonstrate to me how I can join words into whole sentences on a blank page without a cigarette burning away between my lips.' The author John Fowles likewise used the weed as his muse. His writing table was covered in charred grooves where untended cigarettes had burned out.

Both the tobacco industry and its opponents launched assaults on the general public in the 1970s. Each side's attack occurred over the same ground – addiction, or, to be specific, nicotine. The tobacco companies' interest in the compelling powers of nicotine stands in contrast to their research into cancer, which they had effectively abandoned when the subjects of their in-house mice-painting programmes started growing tumours. Tobacco companies had kept a careful watch on nicotine research for decades, commissioning their own reports on any promising developments. These came to the same conclusions – that people smoked cigarettes because of the nicotine they contained, and that nicotine was irresistible. The harder you looked at it, the better it got. As an early and enthusiastic examination of 'The Smoking and Health Problem' by Sir Charles Ellis opined:

> It is my conviction that nicotine is a very remarkable, beneficent drug that both helps the body to resist external stress and also can, as a result, show a pronounced tranquillizing effect. You're all aware of the very great increase in the use of artificial controls,

stimulants, tranquillizers, sleeping pills, and it is a fact that under modern conditions of life people find that they cannot depend just on their subconscious reactions to meet the various environmental strains with which they are confronted. They must have drugs available which they can take when they feel the need. Nicotine is not only a very fine drug, but the technique of administration by smoking has considerable psychological advantages.

Scientists unconnected with the industry were also scrutinizing nicotine, this 'very fine drug', and came up with some equally forthright conclusions as to why people smoked: 'There is little doubt that if it were not for the nicotine in tobacco smoke, people would be little more inclined to smoke than they are to blow bubbles or to light sparklers.' What frightened scientists, and pleased tobacco companies, was how quickly nicotine made itself indispensable to the average human. According to Dr M. A. Hamilton Russell, a pioneering researcher in the field, 'It requires no more than three or four casual cigarettes during adolescence . . . virtually to ensure that a person will eventually become a regular dependent smoker.' This conclusion was a dream come true for the average cigarette manufacturer. The question was how to get the smoke from those three or four casual cigarettes into the lungs of potential customers. Marketing was the obvious answer, but here the tobacco companies ran into a problem: their best potential customers were adolescents, and, strictly speaking, this category was out of bounds.

Tobacco companies had not closed their eyes to all the charges laid against their cigarettes. They too had drawn conclusions from the rise in lung cancer fatalities, and had realized that their clients would suffer from a process they termed 'ageing out', i.e. dying. In a competitive market brand survival depends on numbers. Brands, like religions, cannot self-perpetuate – they always need new converts, or consumers. Cigarette manufacturers realized that the dangers of customers 'ageing out' on them could only be compensated in one way: by targeting the youth market. 'Younger adult smokers are the only source of replacement smokers . . . If younger adults "turn away" from smoking, the industry must decline, just as a population which does not give birth will eventually dwindle.' Doom-laden prognoses such as these, taken from an R.J. Reynolds's document, 'Young Adult Smokers: Strategies and Opportunities', forced the tobacco industry on to the offensive, with the conversion of teenagers into smokers as its objective.

Although it was illegal in most Western nations to sell cigarettes directly to minors, research indicated that it was vital to introduce people to nicotine before they grew up. As a 1977 British American Tobacco document, 'Dependence on Cigarette Smoking – a Review' concluded: 'McKennell (137) has examined data for British smokers and concluded that the onset of regular smoking is an adolescent phenomenon. The 'age of onset' graph looks like a growth curve for other pubertal or maturation phenomena, starting in the early teens. The largest increment in smoking occurs at 15 to 16 years when most people leave school. By the age of 20 most people have already started to smoke if they are ever going to do so at all.' In other words, a cigarette manufacturer's most promising market consisted of 15- and 16-year-olds. The same report also considered why teenagers might be prompted to take the first step towards becoming a customer: 'Russell (168) lists the reasons for the first cigarette as curiosity, conformity, bravado or to appear grown-up. He states that the first few cigarettes are invariably unpleasant and with curiosity satisfied by the first cigarette, smoking is only repeated if physical discomfort is outweighed by psychological or social rewards.'

Tobacco companies, therefore, set out to find out what sort of 'psychological or social' rewards were necessary to encourage young people to start smoking. They conducted discreet market research into the matter – as a firm of consultants advised Philip Morris, the owners of Marlboro: 'True answers on smoking habits might be difficult to elicit in the presence of parents . . . We recommend interviewing young people at summer recreation centers (at beaches, public pools, lakes, etc.).' Research across the industry into the art of tempting minors concurred that the best way of attracting their attention was through sponsorship of sporting events and rock concerts.

While the tobacco companies were busy seeking out sponsorship opportunities whereby they might innocently expose their product to young eyes, scientists were uncovering further evidence on nicotine and its addictive properties that was to serve as ammunition for the anti-tobacco movement in its counter-offensive against cigarette manufacturers. Independent scientific scrutiny was less euphoric in its conclusions on the matter than the tobacco industry. Addiction, especially amongst the young, was not cause for celebration, but a serious social problem.

The analysis of addiction had progressed from psychiatrists and their leather couches to laboratories where biochemists were trying to pin down the actual processes that resulted in humans becoming dangerously obsessed with plant extracts. There had to be more to it than a dreaming Viennese had revealed. They began with substitution trials, to see if people would accept nicotine instead of cigarettes. Substitution trials had been attempted as far back as 1942, when a British anti-smoker, Dr Lennox Johnston, had successfully substituted nicotine injections for smoking, and argued that nicotine and nicotine alone was the reason why people smoked. Regrettably Johnston went insane and plotted to blow up the British Medical Association when it refused to publish his findings, which were thereby discredited. Further trials were carried out in the 1950s and any remaining doubts about nicotine's addictive qualities were ended by a study carried out at the University of Michigan medical school in 1967 that involved injecting subjects with nicotine without telling them when or why, and which proved that people smoked because of nicotine. Even chain-smokers cut their cigarette consumption when they received nicotine from another source.

Independent confirmation that it was nicotine alone, as

opposed to lip eroticism or imaginary child abuse, that caused people to smoke opened the door to other less dangerous ways of administering such a popular drug. Research was carried out into why smoking was such an effective nicotine delivery system. Couldn't a pill or some other slow release mechanism do the same job? What was going on in the smoker's brain, between the neurones? Dr M. A. Hamilton Russell shed light on this problem in a paper which explained that the cigarette was so popular because it allowed the smoker/addict to regulate dosage:

> the level of nicotine in the brain is crucial for the highly dependent smoker. The blood-brain barrier is no barrier to nicotine . . . it is probable that nicotine is present in the brain . . . within a minute or two of beginning to smoke, but by 20-30 minutes after completing the cigarette most of this nicotine has left the brain for other organs . . . this is just about the period when the dependent smoker needs another cigarette.

Here, at last, was the answer to why cigarette smokers seemed to order their lives around their habit – they needed another fix every half hour or so. This was also why cigarette smoking had advanced in lock-step with mass production and the consequent standardization of working hours. People had little time for smoking, and the instant hit a cigarette offered could be accommodated within the working day. In contrast, pipe and cigar smokers who did not inhale had to wait more time for the nicotine to kick in, but enjoyed their high for longer. There was, therefore, some truth in the settled character attributed to pipe smokers and the aura of relaxation that surrounded their nicotine delivery system. It was incapable of offering a cigarette-like hit, but with patience and sufficient hours of leisure would provide a smoother, more enduring trip.

The sum of scientific investigation into nicotine in the 1970s was that the substance was so addictive that even a very small amount could turn a person into its slave, and that the behaviour of smokers could be explained in its entirety in terms of chemical dependence. There was, however, a flaw in this model of smoking. Research had revealed that some smokers were more addicted than others. Not every smoker lost control of their tempers and/or bowels when deprived of cigarettes for more than twenty-four hours. A proportion of them appeared to be indifferent to nicotine's presence or absence in their bodies. In other words, plenty of smokers were smoking for the pleasure of the act alone. This was unpleasant news for any scientist hoping to formulate a universal theory of smoking. It was so much easier to categorize all smokers as addicts, i.e. victims. This enigma was solved by recourse to Freud. Even Dr Hamilton Russell could not resist a dash of Oedipus complex. He identified five types of smokers, coincidentally the same number of them as humours, whose different styles of addiction manifested themselves in their characters. The five types were: (1) 'psychosocial' smokers who were weak addicts – the sort of people who smoke at parties or after one night stands; (2) indulgent smokers, who used cigarettes for sensual or sensory rewards; (3) self tranquillizers (Freud's lip eroticists); (4) self-stimulators, who used cigarettes as a drug to keep them awake or to help themselves to concentrate; and (5) addictive smokers who were simple, helpless addicts and in desperate need of protection.

Once the 'helpless victim' model of nicotine addiction had been perfected by turning smokers who were not physically addicted into psychological addicts, tobacco's opponents were able to claim that smokers did not need to be wicked, merely careless to end up with a potentially fatal drug habit. They promptly attacked the tobacco manufacturers, labelling them slavers in addition to merchants of death, and they began to pressurize governments into taking firmer action against cigarette smoking. Although warning labels and advertising bans had had an effect on the number of adult smokers, children either ignored them, or welcomed them as a challenge. By the 1970s it was impossible to find anyone in the Western world who had not at least heard of the dangers of smoking, but it seemed that wise words were not enough. Tobacco's opponents began to lobby for restrictive measures not just against the tobacco companies, but smokers too.

At this point, the smoking debate had arrived at a philo-sophical crossroads. Both the British and American govern-ments were holding back from taking further measures against their subjects and citizens. They had been warned, regulated and taxed. It would be a breach of the social contract to impose further on individual rights. Cigarettes killed people, but so did cars, and, for that matter, eating too much. The fundamental liberal principle of democratic government – that state intervention should be limited to occasions when an individual's behaviour might damage others, and not if they only risked themselves – stood firm in the Anglo-Saxon world.

It did, however, come under heavy pressure from a variety of organizations opposed to smoking, which had become better organized and better financed as the 1970s progressed. By the end of the decade, people were finding there was money and power to be had out of opposing tobacco. It was possible, for instance, to hold the tobacco industry up to ransom. The British government had long been expert at this, waving a casualty register with one hand and extending the other for more money. The tax rises it introduced in 1981 finally pushed the price of a packet of cigarettes over the psychological barrier of £1.

The tobacco war was also an important source of revenue for the legal profession, which adopted the role usually reserved for profiteers in a conflict. Lawyers were active in both camps, fighting for and against the right to smoke. American cigarette manufacturers had a crack battalion of combat seasoned advocates on duty at all times, who could spin out a cancer case until the plaintiff was in the ground and his relatives too impoverished to continue, or testify before the US Congress of the good that tobacco did for America, enumerating the multitude it employed, the mouths it fed, and the billions it earned, to the nearest cent.

Tobacco also enriched the world of science. Epidemiologists were the first of its practitioners to find fame and wealth on the tobacco battlefield. Dr Ernst Wynder, whose 1950 studies and later adventures with mice and Lucky Strikes had catapulted him to stardom in the scientific community, was driving a sports car and dating actresses by the 1970s. As the refugee from Hitler put it: 'I've had the two best things you can hope for in life, a German education and American opportunity.' Epidemiologists were soon joined by other experts at the trough. Medical science, which had been twiddling its thumbs since typhus was slain, at last had an epidemic worthy of its attention, whose suppression offered opportunities for both enrichment and glory. Felicitously, smoking was a useful culprit for many fatal ailments which killed without betraying the secret of their cause. Heart disease, emphysema, gastric ulcers and chronic bronchitis could all be blamed on smoking with varying degrees of accuracy.

Smoking was a particularly useful excuse for getting grants for research into heart disease. People no longer died of broken hearts but heart attacks, which needed to be explained in more definite terms than unrequited love on a death certificate. Smoking had been an accepted statistical cause of heart disease since the first Royal College of Physicians report in 1962. Unlike lung cancer, the mechanism by which cigarette smoking caused heart attacks was easier to prove. Nicotine speeds the heart beat and 'the candle that burns twice as bright burns half as long'. Every mammal's heart, from shrew to elephant, beats roughly the same number of times before surrender, and so unnecessary exertion shortens lifespan, which is why smokers, athletes and the obese tend to die prematurely.

Meanwhile, on the other side of the battlefield, tobacco company money was always available for serious research, if sometimes into frivolous matters. Many scientists were happy to accept its gold, some of whom believed that by working with the Devil they might tame him. Unfortunately for such idealists, cigarette manufacturers were most interested in the science of appearances. Here is one of the fruits of tobacco industry sponsored research, explaining why the smoke from a burning cigarette looks so tantalizing:

> Smoke rises from a cigarette because it is hot and lighter than air. The stream is smooth and shaped like a column because it is laminar, and all the smoke particles travel in parallel.
>
> The flow is laminar because the cigarette's burning area is small, and the energy output that is driving the stream upward is only about one watt. Larger sources of smoke such as smokestacks or bonfires clearly do not produce laminar streams.
>
> The scattering of the smoke marks the transition of the flow from laminar to turbulent, and is caused by the growth of small, invisible disturbances to the smoke stream that eventually create scattering. Turbulent flow is characterized by random fluctuations of speed and direction.
>
> Careful measurements using lasers and special imaging techniques have revealed that the smoke stream is only a small part of a much larger invisible plume that rises from the cigarette, with the hottest and fastest moving part of the plume actually a few millimeters in front of the smoke stream.
>
> As the plume rises, it takes the surrounding air with it and cools the air down which, surprisingly, makes it move faster.
>
> In very still air, I have observed stable plumes that are as long as 30 centimeters, which become wavy just prior to scattering.
>
> PETER LIPOWICZ
> Senior principal scientist
> Philip Morris USA
> Richmond
> Virginia

The tobacco war was fought in American courtrooms, in Britain's Parliament, over the allocation of seats on trans-atlantic flights, in the newspapers, whose war reporters spilled more successive ink each year – tobacco reporting became a surefire way to pay a mortgage – and at every point of contact the conflict generated cash. It was a global contest, with endless opportunities for glory. The net result was that a large body of people gained a professional interest in tobacco. It was no longer simply a matter of companies selling it and customers buying – there was plenty of inter-vention work available in the middle.

As particle physicists were at that time postulating, matter suggests anti-matter, and the organizations and individuals dedicated or paid to suppress smoking – the anti-tobacco movement – adopted all of the tobacco industry's policies, with opposite aims. It began to indoctrinate youth, to falsify statistics – or to ignore them when they were inconvenient – and to advertise in all the principal media. The anti-tobacco industry's critical initial failure as a true invert was to present death as anti-glamour. Young people admired the health warnings on cigarette packets – they were like initiation scars, or battle honours – part of the rites of passage of growing up. As a consequence, the anti-tobacco movement had little success against teenage smokers. Adults proved to be more receptive to the anti-smoking campaign, and were more easily influenced by mortality statistics. By the begin-ning of the 1980s they were even trying to give up Silk Cut. In 1982, the UK's percentage of adult male smokers had fallen below 40 per cent and the percentage of female smokers had dropped to around 35 per cent.

The decline was a significant advance for the anti-smoking cause. The question was how to turn advance into total victory and consign smoking to the history books. Irritatingly, remaining smokers were protected by an impregnable barrier named 'freedom of choice'. People chose to smoke tobacco, and it was legal all over the world. Smoking was a victimless holocaust, because the dead people had chosen to die that way. Or had they? Was not

18

Man and Superman

The tobacco habits of the underclass – the quest for the safe cigarette – can smoking kill non-smokers? – tobacco buys into Hollywood – tobacco companies fall victim to predators on Wall Street – Western brands penetrate the Eastern Bloc – cigars, power and glamour

As the Americans had observed, a majority of the remaining British smokers were poor, or 'working class' to use the quaint British terminology, and the working class were suffering their worst decade for a long time in the 1980s. As Great Britain's economy made the transition from industrial to service orientated, many of the working class lost their work, and the fallen were numerous enough to merit a new social stratification – the underclass. In principle, the underclass relied on the state to pay their rent, their medical bills and to provide them with pocket money, much of which the state reclaimed immediately in cigarette tax. Unemployment benefit did not keep up with cigarette tax inflation, resulting in the birth of discount brands of cigarettes. The venerable name of Lambert & Butler was resurrected and attached to a cut price make of cigarettes, whose strident black and grey packaging did nothing to evoke the luxury and elegance once associated with the purveyors of the best cigars in London.

Despite the assault on their pockets, the members of society's poorest strata remained faithful to cigarettes, giving up in far fewer numbers than their counterparts in the professional caste. They kept this devotion in the face of ever larger warnings on cigarette packets, and without advertising to inspire them, as cigarette adverts, under one of the many voluntary agreements reached between manufacturers and the British government, could no longer show humans and cigarettes together.

The new underclass were castigated for smoking by social workers and were bombarded with tobacco fatality statistics. They were told that filter cigarettes, including the ultra low tar and nicotine cigarettes were no safer than their parents' gaspers, even though the average cigarette smoked in 1980 yielded 14 mg tar and less than 1 mg nicotine, against 37 mg tar and 2 mg nicotine in 1965. These figures, however, did not tell the whole truth as a practised smoker could extract more from his cigarette than any government testing machine. Indeed, certain design features in the low tar brands seemed to have been engineered to allow the smoker nicotine and tar doses in excess of government ratings. The most obvious of such features were the ventilation holes in the middle of filters, precisely positioned so that they would be closed by a smoker's fingers. With concentration and no more than average lung power, a light cigarette smoker could get up to six times his cigarette's official yield. Such behaviour was termed 'compensation', and was symptomatic of the frustration smokers faced from having to rely on weak cigarettes which they were told would kill them anyway. According to the British government, the only safe cigarette was the one you did not smoke.

The same held true across the Atlantic, although the US tobacco industry had tried to discover the smoking world's equivalent to Eldorado – the safe cigarette. Between 1977 and 1979 the American manufacturer Liggett & Myers had worked on a new type of cigarette, to be branded 'Epic', which used palladium to catalyse the cigarette smoke before it entered the smoker. Mice painted with palladium catalysed residues did not grow tumours. But the results of these tests with mice could not be used to market Epic because of the new rules governing cigarette promotion. The most forward advert mocked up for the new brand advised that Epic was 'worth trying just for the taste but – UNTIL THE GOVERNMENT CHANGES ITS POLICIES, WE CAN'T REALLY TELL YOU WHAT'S *NEW* ABOUT EPIC'. Given that people were on their guard against cigarette advertising, the copy was weak for what might have been a revolutionary product. Further, the problem with selling a safe cigarette was the same the manufacturers had encountered forty years previously when they were selling healthy cigarettes: if one sort was safe, the rest were not. L&M's lawyers were utterly opposed to the idea, and Epic's lifespan did not live up to the promise of its name. The product was killed before it reached the market.

In the absence of a safe cigarette, the question on every smoker's lips was 'is there a safe limit to smoking unsafe cigarettes?' For instance, was it safe to smoke ten cigarettes a day, and if not what about five, or even just one for breakfast? And if there was no safe limit, nor even a quantifiable diminishment of risk from smoking fewer, weaker cigarettes, then why not trade up to high tar and smoke as much as you could afford?

Surprisingly little research had been carried out into establishing a safe limit for smoking cigarettes. The manufacturers' lawyers advised them that admission of a safe limit was tantamount to admitting that cigarettes could be unsafe in excess, whereas smoking's opponents could not allow any cigarette to be safe, nor any multiple thereof. The neglected smokers in the middle felt there had to be a safe limit. After all, people had smoked for centuries without dying of cancer, indeed, the current life expectancy of pipe smokers was not greatly different from that of non-smokers. Tobacco, therefore, did not necessarily kill, and if you could smoke a pipe all your life without needing to worry, then why not a few cigarettes every day?

Any attempts to solve this enigma were abandoned when

the anti-smoking movement attacked cigarettes from an unexpected angle. Not only were cigarettes unsafe to smoke, but they were also unsafe not to smoke, i.e. to non-smokers. Even the ones you did not smoke might kill you. This counter-intuitive revelation resulted in something akin to a gold rush in the legal and scientific communities. From a legal point of view it was a dream come true. If smoking killed innocent bystanders every victim would need a lawyer and be worth millions in compensation. In the eyes of the scientific community it represented a convenient way to bypass the flaws in the addict-as-victim model. If smokers really were killing non-smokers with their habit then they must be quarantined and treated at once.

The issue of passive smoking was born. It was christened 'Environmental Tobacco Smoke' or ETS for short. Its pedigree, in the sense that it might possess a lethal potential, stretched back to 1963, when the term passive smoking was coined as '*Passiver Zigarettenrauchbeatmung*' in a study that examined tobacco smoke's potential as an environmental pollutant. It was considered for this role once again in the 6th US surgeon general's report of 1972. Passive smoking made its debut on a wider stage at the World Health Organization's third world conference on smoking and health in 1975, where it appeared dressed as a wish, and was announced by a speaker at the conference with the following lines: 'We must foster an atmosphere where it is perceived that active smokers would injure those around them, especially their family and any infants or young children.'

ETS next appeared in 1981 in Japan in a study carried out by Dr Hirayama, published in the *British Medical Journal*, on lung cancer rates among the non-smoking wives of smokers. Japanese smoking habits were ideally suited to the study, as 65 per cent of Japanese men smoked, and less than 10 per cent of the country's women. The only technical problem the survey faced was that very few Japanese of either sex died of lung cancer. Nevertheless, Dr Hirayama succeeded in establishing a statistical link between ETS and lung cancer in the non-smoking wives of smoking husbands.

Coincidentally, his report also demonstrated that these non-smoking wives were more likely to get lung cancer than wives who did smoke. The revelation that not smoking killed non-smokers was neglected, and the slender evidence Hirayama offered of ETS's lethal potential was embraced. In the anti-smoking movement's eyes it was the first step towards the discovery of their Holy Grail – proof that smoking killed innocent bystanders. Interestingly, the report was criticized by the head of Hirayama's research institute who pointed out that Japanese men were not at home often enough to expose their wives to much smoke.

These tentative beginnings led to a crusade as scientists everywhere sought and were awarded grants to pin various diseases on to passive smoking. The cross was even assumed by the then US surgeon general, Dr Everett Koop, a veteran campaigner against tobacco, whose aim was a 'Smoke-free America by the year 2000.' In addition to harbouring millennial ambitions, Dr Koop put his name to the 19th US surgeon general's report of 1986 which stated in its preface: 'It is certain that a substantial proportion of the lung cancers that occur in nonsmokers are due to ETS exposure.' The assertion was so qualified in the body of the report as to question its validity, and the overtly political aims of a purportedly medical document were described by a congressman from a tobacco growing state as 'a very deliberate attempt to turn nonsmokers into antismokers'.

Meanwhile, numerous epidemiological and other studies had been launched to search for links between ETS and dead non-smokers. Few had any success, some even concluding that passive smoking was good for non-smokers. While ETS was certainly an inconvenience to non-smokers, who suffered the smell and unpleasantness of tobacco smoke without any of its addictive pleasures, the suggestion that it was a killer was at best unproven. The position on passive smoking at the end of the decade was summed up by Stanton Glantz, a leading light in the Californian non-smoking movement: 'The main thing that science has done in addition to help people like me to pay mortgages, is it has legitimized the concerns that people have that they don't like cigarette smoke. And that is a strong emotional force that needs to be harnessed and used. We're on a roll and the bastards are on the run.'

There is, however, one area where passive smoking is a confirmed killer. For decades scientists had been fascinated by the phenomenon of spontaneous combustion. Human bodies are occasionally discovered reduced to a pile of ashes, having burned at a heat greater than the average crematorium, which merely burns the flesh off bodies (the skeletons are pulverised in a separate operation then added to the urn given to relatives). It seemed that every victim of spontaneous combustion was a smoker, and after experiments carried out with pigs' bodies wrapped in rags, the answer to the mystery appeared to be that smokers had fallen asleep while smoking and the cigarette had done the rest of the work.

The tobacco industry reacted coolly to the ETS issue. It was already under assault from all quarters and any non-smoker who actually claimed to be dying of smoking would need spectacular circumstantial evidence to get their case to trial. The industry's hands were already full of dying smokers and customers took precedence. It still had a 100 per cent record in litigation. No cigarette manufacturer had yet paid any compensation to any smoker anywhere in the world. The tobacco industry had made many lawyers wealthy, but these were doing a good job, and if an injured smoker's chances were bleak, then those of a non-smoker were non-existent.

The tobacco industry got on with what it did best – providing people with reasons why to smoke. Since television was out of bounds, it concentrated its efforts on feature films, and if smoking was not already scripted into a movie, its presence could be purchased. The process of paying film-makers money to feature branded goods was called 'product placement'. Philip Morris, as usual the leader in marketing innovations, set a gold standard for product placement in *Superman II* (1980). For a reported payment of $42,000

Marlboro appeared twenty-two times in the film, whose climactic battle between the Man of Steel and various foes takes place against a backdrop of Marlboro billboards. Lois Lane, Superman's girlfriend, smokes Marlboro throughout the film, a habit she had never displayed in fifty years of comic book appearances.

As the 1980s progressed, cigarette manufacturers became fervent product placers. Hollywood welcomed them with open arms – after all, people smoked in real life, so what was the harm of letting them do it in the movies? Smokers in the audience would not be fooled by fictional brands, so why not film actors smoking real ones and make some money on the side? As the following example shows, both the film and tobacco industries considered product placement to be lucrative and inoffensive:

Sylvester Stallone

April 28, 1983

Mr. Bob Kovoloff
ASSOCIATED FILM PROMOTION
10100 Santa Monica Blvd.
Los Angeles, CA 90067

Dear Bob:

As discussed, I guarantee that I will use Brown & Williamson tobacco products in no less than five feature films.

It is my understanding that Brown & Williamson will pay a fee of $500,000.00.

Hoping to hear from you soon;

Sincerely,

Sylvester Stallone
SS/sp

(M)

**ASSOCIATED
FILM
PROMOTIONS**
An AFP, Inc. Company

June 14, 1983

Mr. Sylvester Stallone
1570 Amalfi Drive
Pacific Palisades, CA 90272

Dear Mr. Stallone:

In furtherance of the agreements reached between yourself and Associated Film
Promotions, Inc. representing their client Brown & Williamson Tobacco Corp. (B &
W), I wish to put in summary form the various understandings and details regarding
B & W's appearances and usage in your next five scheduled motion pictures. B & W
is very pleased to become associated with the following schedule of films and to have
you incorporate personal usage for all films other than the character of Rocky Balboa
in Rocky IV, where other leads will have product usage, as well as the appearance
of signage (potentially ring).

The following is the current list of the next five (5) minimum films for B & W's
appearance. It is understood that if production committments change the order or
appearance of any of the group of films to be released, B & W will appear in a substituted
film. The only non-appearance for B & W will be by mutual consent of both parties,
in which case another Sylvester Stallone movie will be arranged for substitution.

The initial schedule of films is:

A). Rhinestone Cowboy   D). 50/50
B). Godfather III   E). Rocky IV
C). Rambo

In consideration for these extensive film appearances of B & W products, Brown and
Williamson agrees to forward to Robert Kovoloff and Associated Film Promotions,
Inc. their initial deposit to you of Two-Hundred-Fifty-Thousand Dollars ($250,000.00).
This represents a fifty percent (50%) deposit of the total financial committment by
B & W. The subsequent Two-Hundred-Fifty-Thousand Dollars ($250,000.00) is agreed
to be forwarded in five (5) equal payments of Fifty-Thousand Dollars ($50,000.00) each
payable at the inception of production of each participating film.

On behalf of our client Brown & Williamson Tobacco Corp., we wish to thank you for
this long term committment, and look forward to each release from the excellent
schedule of films that they will participate in.

Very truly yours,

James F. Ripslinger

James F. Ripslinger
Senior Vice President

JFR:jag

cc: James Coleman, Brown & Williamson Tobacco Corp.

(M)

2401.15

and cats, was used to transform the distant, staid bactrian on the old Camel packet to a mutant named 'Joe Camel', whose smiling face dominated the new box. Joe Camel did not have an easy birth. His corporate parents had to learn to love him. The naked woman no longer was visible on his foreleg, but a giant penis had appeared between his eyes. The Company tampered with the image, but stuck with the phallic version when advised 'That's what Camels really look like.'

In autumn of the same year as Joe Camel's birth, Ross Johnson, ex-CEO of Nabisco who was tired of the tobacco spectre dragging down the conglomerate's share price, announced to the RJR Nabisco board that he intended to lead a management buy-out, and purchase the company for $17 billion. The announcement had the same effect as spilling blood in front of sharks. The HLT specialists began to circle and RJR Nabisco was eventually taken private by KKR in 1989, for a record $24.9 billion – then the largest corporate transaction of the twentieth century. It was asset stripped at breathtaking speed, and unpromising product lines, like 'Premier', a smokeless cigarette, were dumped. Joe Camel was heavily promoted and a survey published in 1991 showed that 91 per cent of six-year-olds could match Joe Camel to his product (cigarettes), and that the nasally disadvantaged bactrian was as recognizable to schoolchildren as Mickey Mouse.

One of the factors that had enabled the HLT specialists to value a tobacco company more highly than the stock market was the importance of overseas sales. By the late 1980s the American tobacco companies were true multinationals, with much of their earnings deriving from foreign sales. Their export drive had been assisted by the US government which had sometimes used tobacco as a lever to open foreign markets to US goods. Japan, in particular, was a sore point. It enjoyed a huge trade surplus with America, which had become an important political issue.

Senator Jesse Helms wrote to the Japanese prime minister in 1986, reminding him of America's defence commitments in Asia and explaining he and his colleagues in Congress 'Will have a better chance to stem the tide of anti-Japanese trade sentiment if and when they can site tangible examples of your doors being opened to American products.' The example Senator Helms had in mind was cigarettes: 'I urge that you establish a timetable for allowing US cigarettes a specific share of your market.' When the Japanese complied and removed tariff barriers in their market against foreign cigarettes consumption leapt, especially among the young. Despite the much vaunted cultural differences between the countries, it seemed that teenagers everywhere were the same, and the US cigarette companies had two decades' experience of selling to underage smokers. A similar sequence of events occurred in Taiwan, whose market had hitherto been dominated by a domestic brand auspiciously named 'Long Life'. Within a few years of American brands being allowed into the market, cigarette consumption amongst schoolchildren had risen by over 50 per cent.

The 1990s offered even more opportunities for overseas sales, especially in the Eastern Bloc, which represented a particularly promising market for cigarettes. Most comrades smoked, and did so with the same innocent enthusiasm that had characterized Westerners before they started dying in droves from lung cancer, heart disease, emphysema, etc. Having little opportunity to aspire to any more sophisticated form of consumerism, Russians aspired to smoke more, and as befits the first nation in space, they were also the first to take a smoking implement into orbit. A specially made Cuban cigar travelled aboard Soyuz 30 in 1980. Not merely content with carrying tobacco off planet, the USSR also provided the first man to smoke (and drink vodka) in space. Valery Ryamin, Soviet cosmonaut, took a cigarette with him on Salyut 6, and returned without it. When questioned on his off-world activities, Ryamin's response was typically Russian, and beautifully enigmatic: 'all sorts of experiments may occur in the environment'.

Throughout the Cold War, few Western cigarette brands had penetrated the USSR and its satellite states. Marlboro had made the greatest progress – it had been sold in the special Freedom shops (reserved for foreigners) since the seventies, and had had a single, if spectacularly successful publicity coup when the Soviet leader Leonid Brezhnev held up a carton of Marlboro, specially packaged to commemorate the Apollo-Soyuz space mission, on Russian television.

However, by the end of the 1980s, the comrades had become disenchanted with the shortages and wars that seemed to be a consequence of the communist system. Russians were bored with queuing up to buy bread, or an onion, and they no longer believed that life was the same, or worse, in the West. Soldiers returning from the war in Afghanistan told of the relative riches of a Third World economy in a battle-zone, and the comrades began to dream of smoking Marlboros and driving Mercedes. Further, the crumbling Soviet system could no longer keep Russians supplied with state manufactured cigarettes, which led to protests in Moscow in August 1990. Realizing his glasnost programme might lose its authority if the workers could not be supplied with something as simple and essential as cigarettes, the last Soviet leader Mikhail Gorbachev was forced to issue emergency orders to American manufacturers for billions of cigarettes, which were paid for with diamonds, oil and gold.

When the Berlin Wall fell and the USSR dissolved into a dozen separate states, the tobacco multinationals moved into these new and promising markets. In many cases they established joint ventures with the old state manufacturers, but when local tastes and investment policies permitted, they manufactured their Western brands and encouraged their importation, so that these tokens of unrestrained capitalism were soon common in the heartlands of communist ideology.

If viewed from a global perspective, mankind's love affair with tobacco blossomed in the late 1980s. The notion that cigarettes killed and the Western concept of addiction were unknown to much of the planet's population. Tobacco production was increasing, and for every white man who died of lung cancer there were ten Asians to replace him.

Even in Western societies, where the cigarette had been labelled the biggest killer in history (pace Lenin and Chairman Mao), other forms of tobacco consumption flourished. The American preference for eating tobacco enjoyed something of a revival during the decade. It had at least 12 million aficionados, concentrated in rural areas where spitting was still acceptable, and was paraded before a wider audience by America's major league baseball stars, a third of whom used some form of smokeless tobacco, taking it up their noses as well as in their mouths. Chewing tobacco was promoted as the athletes' choice, even on television, where its advertising was still permitted, and prospective chewers were advised: 'in two weeks you'll be a pro'. Although the outlook for chewing tobacco was not entirely cloudless – a schoolboy track-star addicted to chewing had died publicly and painfully of tongue cancer and his parents had launched a suit against the manufacturer of his favourite brand – it was considered promising enough for United States Tobacco, king of the chewing market, to launch 'Skoal Bandits', a kind of tobacco tea-bag which the dedicated lover of nicotine might keep tucked under their upper lip in the Swedish fashion.

Meanwhile in Great Britain the traditional art of pipe smoking continued to attract new adherents. While in gentle decline overall, the habit did not attract the vilification accorded to cigarettes, and indeed was considered sufficiently respectable for politicians to dare to appear in public smoking a pipe. Neil Kinnock, leader of Great Britain's Labour opposition, and Tony Benn, a member of his shadow cabinet, were both committed pipe-men. Pipe smoking was patronized by a broad cross-section of society. Winners of the British 'Pipesmoker of the Year' awards included the disc jockey Dave Lee Travis in 1982, the astronomer Patrick Moore in 1983, the ex-heavyweight boxing champion Henry Cooper in 1984 and the cricketer Ian Botham in 1988.

The 1980s also witnessed the birth of an entirely new form of tobacco consumption which, in keeping with the trends towards 'lite' foods, pale beer and white spirits such as vodka over the traditional staples of red meat and whisky, succeeded in reducing the weed to its essence. While the tobacco companies had been diversifying out of tobacco and focusing on overseas cigarette sales, other manufacturers were diversifying into the tobacco market. Drug companies, in particular, were attracted by the profitability of nicotine addiction. There were no expensive patents attached to the narcotic and people were willing to pay more for a packet of cigarettes than a bottle of aspirin. Nicotine users were also highly predictable – factories could be planned around them if they liked a brand. The challenge was to find a delivery system that: (a) did not involve smoking; (b) could be disguised as a health product; and, possibly, (c) was 'stimulating, pleasurable and flavorful'.

The first non-tobacco company entry into the tobacco market was Nicorette, a brand of nicotine impregnated chewing gum. The colour of Nicorette's packaging was surprisingly similar to that of Kool, the original mentholated brand, and its flavouring almost identical. Despite the knowledge that chewing gum ruined people's teeth and gave them jowls and stomach ulcers, and without commenting on the habit's effects on its users' manners, chewing gum was perceived to be a healthy occupation, especially suitable for children. Further, nicotine gum could be marketed as an aid to giving up smoking. In principle, a smoker could chew gum instead of lighting up, and once they had renounced cigarettes, they could proceed to wean themselves off the gum.

Surprisingly, the rationale behind nicotine gum appeared to contradict the anti-tobacco movement's assertion that nicotine was terminally addictive and that the quantity contained in no more than four cigarettes could hook an adolescent for life. (Incidentally, the 'helpless addict' approach to nicotine had been adopted by the US surgeon general in his 1988 report, the twentieth in the smoking series.) What if children got hold of nicotine gum? A Nicorette addiction would make short work of their pocket money. Nicorettes came with a health premium in their pricing, presumably to reflect the belief that people would follow the product's instructions and give them up as soon as their work as an aid to quitting smoking was done. Sensibly, their sale was limited to chemists, to prevent them falling into the wrong hands. Nicorettes soon found a market among diehard smokers who used the gum as a supplement to smoking, or to carry them through times when they could not smoke, such as long distance flights, on which many airlines had banned smoking altogether.

Smoking bans had proliferated throughout the decade, and by its end the old-fashioned presumption that smoking was permissible unless specifically prohibited had been reversed. Most workplaces, public places and forms of public transport had forbidden smoking. Not all such bans were resented by smokers. When London Underground finally prohibited smoking on its trains and platforms in 1989, London lost one of its most sordid spectacles and perhaps the last vestige of Victoriana in the capital – the smoking carriage – a 20-yard ashtray filled end to end with tobacco smoke. The cigarette was also displaced from its government approved role as the soldier's friend. Cigarettes had long been absent from GIs combat rations, and when America sent a task force code-named 'Desert Shield' to the Persian Gulf in 1990 to protect the oil-fields of Saudi Arabia from Saddam Hussein, the troops had to buy their own cigarettes. The tobacco companies attempted to reassume the role they had played in previous wars. Before the Pentagon prohibited the practice, Philip Morris had donated over 2 million cigarettes to members of the US armed forces. Unfortunately for the manufacturers, American concern for its servicemen's lives led to a ban not only on free cigarettes but also on all forms of battlefield tobacco promotion. For instance, a shipment of 200,000 magazines to Desert Shield personnel was held up while special wrappers advertising Camel filters were removed.

While the 1980s represented the most successful decade in the anti-smoking movement's history – people exposed to the health message had begun to listen, and non-smokers could at last be sure of protection at work and in public places from the irritation and odours of other people's tobacco habits – their victory in the West was far from complete. Cigarette smoking had declined, but cigar smoking had increased. The same political factors that had created the underclass in Britain and the trade deficit in the United States had bestowed benefits as well as burdens. The richest fraction of society in each country had become comparatively richer and some of their excess wealth was devoted to purchasing cigars, preferably large ones. The cigar divans of nineteenth-century London were recreated in New York and Los Angeles, and at the same time as the media available for cigarette advertising were diminishing and No Smoking signs had become as common as those stating Exit, a new magazine, entitled *Cigar Aficionado*, was launched, dedicated to the pleasures of owning and smoking the most expensive and flamboyant of smoking devices. Surging demand in the Anglo-Saxon world led to a corresponding increase in the production of quality cigars. The principal beneficiary was the Dominican Republic, whose output leapt by 18 per cent to 55 million cigars in 1993.

The cigar revival is explained in part by the spirit of conspicuous consumption which permeated the age. Cigars were a field in which discrimination and purchasing power could be exercised with equal freedom, and smoking a cigar was a perfect demonstration of both. A new answer to the question 'Why smoke?' was thus added to the extensive list: to prove wealth. Curiously, the most sought after cigar for this purpose was the Cohiba – the brand created for Fidel Castro as a symbol of communist supremacy. When the USSR fragmented, aid payments to Cuba ceased and these treasures were released for sale to capitalists to gain much needed foreign exchange, thus allowing Cohibas to become success symbols in two opposing political systems. The Cohiba became the definitive 'power smoke', and the term 'power smoking', denoting the flamboyant consumption by someone successful or wealthy of a large cigar, entered the vernacular.

Cigars were not only used to express wealth, but to win it. As the great Prussian statesman Bismarck had observed in 1871:

a cigar held in the hand and nursed with care serves, in a measure, to keep our gestures under control. Besides, it acts as a mild sedative without in any way impairing our mental faculties. A cigar is a sort of diversion: as the blue smoke curls upwards the eye involuntarily follows it; the effect is soothing, one feels better tempered, and more inclined to make concessions – and to be continually making mutual concessions is what we diplomats live on.

Material girl

While few concessions have been traced to Bismarck, the utility of a cigar when bargaining remained unchanged, and they were popular on Wall Street and in the City of London during face-to-face negotiations on the giant corporate transactions so common in the age.

This time round the cigar was a power symbol for women as well as men. Actresses as well as merchant bankers took up a cigar habit, thus reviving a western American custom first noted by travellers to Spanish Santa Fe at the turn of the nineteenth century. This phenomenon was a reflection of the spirit of equality that had permeated Hollywood during the decade, in which actresses improved their relative standing, leading and directing films as well as making eyes at leading men. The cigar had always been a favourite with male directors, and now it became popular with actresses and other independent women in the entertainment industry. The new fashion attracted much media comment, which assisted in popularizing the habit amongst women in other occupations.

Surprisingly, cigar smoking did not attract the same opprobrium as cigarette smoking. It was considered to be a less pressing problem than the cigarette epidemic, and its victims (if any) could be presumed to be capable of paying their own medical costs. Further, cigars were not the product of death-dealing multinationals, but the result of the manual labour of otherwise disadvantaged workers in the Third World and therefore, possibly, politically correct. An example of this conundrum appeared when William Jefferson Clinton ascended to the White House as the 42nd president of the United States of America in 1993. Cigarettes were promptly banned but cigars, as will be shown, still had their place in America's seat of power.

19

Hunde, Wolt Ihr Ewig Leben?

*The propaganda battle between cigarette
manufacturers and the opponents of
smoking continues – Marlboro Man rides
again – smoking and supermodels – death
and taxes – legal victories against the
tobacco industry – tobacco smuggling – the
future of the affair between mankind and
La Diva Nicotina*

Tobacco's progress, and that of its detractors, in the most
recent decade of the weed's history can be summarized as
'the oldest sins the newest kind of ways'. The 1990s
witnessed a growth in restrictions against tobacco, combined
with a decline in smoking in the industrialized West, and an
increase in tobacco consumption on a global scale. Turning
first to the situation in the West, the anti-smoking movement achieved the upper hand politically over the course of
the decade, although it met with differing success in its two
principal aims – to cure the world of its tobacco addiction,
and to restrict the rights of smokers.

It has advanced least in the first field. No cure has been
found for lung cancer, although treatment methods have
greatly improved and survival rates for victims have
increased. This is a matter of concern for non-smokers as
much as smokers: putting an end to smoking will not put
an end to lung cancer. However, heartening evidence of the

effect of quitting smoking on lung cancer has emerged from California. Using data gathered between 1988 and 1997, during which time the State spent $634 million (part funded by tobacco taxes) on tobacco use reduction efforts, lung cancer rates in California declined by almost 13 per cent.

Tobacco's opponents have also made little or no progress in their attempt to prove smoking kills non-smokers. Although ETS (passive smoking) is commonly declared to cause an increased risk of lung cancer, in particular in Great Britain, reliable studies have not found this to be so. The American Cancer Society's massive cohort study of 250,000 people published in 1997 did not find a significant relation between passive smoking and lung cancer. The World Health Organization's principle study of the subject, whose delayed publication in 2000 led to intense conjecture that the hesitation resulted from political motives, again found no significant link.

Bearing the foregoing in mind it is curious that the British government's Scientific Committee on Tobacco and Health (SCOTH) stated as a 'Key Message' in its 1999 report that 'Passive smoking is a cause of lung cancer and childhood respiratory disease. There is also evidence that passive smoking is a cause of ischaemic heart disease and cot death.' This conclusion was achieved in part by discarding reports with contradictory conclusions, and partly by resorting to unsound statistical analysis. SCOTH's conclusion regarding passive smoking as a cause of cot death appears to be no more than a fabrication. The alarming rise in the number of cot deaths in the early 1990s coincided with a drop in smoking and a proliferation in smoke-free zones. Cot death was almost unknown when smoking rates were much higher, and the surge that occurred in the 1990s has been attributed to a British government policy of advising mothers to put their babies to sleep on their stomachs, which has since been revoked, with a corresponding decline in casualties. Passive smoking does not cause cot deaths, and with the possible exception of people who spent their lives in smoke saturated environments, such as the comedian and musician

Roy Castle, who died of lung cancer after a career spent in the polluted air of bars and night clubs, there is no proof of it killing anyone at all.

The propaganda battle between cigarette manufacturers and opponents of smoking was prosecuted enthusiastically during the 1990s, pitting cunning against common sense. Although the tobacco companies were increasingly circumscribed in their ability to fight propaganda with their usual weapon of advertising, they adapted to the times, proving ever more adept at reading the public's moods and placing their brand names before their eyes. This was achieved in part by 'brand stretching', which involves using a cigarette brand for clothing for example, thus keeping the name and its associations with style visible. Both the Marlboro and the Camel brands were attached to leisurewear lines in the course of the decade. Marlboro's own assessment of Marlboro Man's place in the caring 1990s shows how clearly the company understood its market. An internal document notes: 'The emergence of a new trend we call "Wildering".' Wildering is defined as 'A new appreciation of ourselves as primal beings, returning to nature to test ourselves, physically and spiritually,' and 'demonstrates that the impulse that drives the cowboy myth has never been more compelling to consumers.'

The Wildering phenomenon required changes to be made to Marlboro's cowboy to bring him up to speed with the times. A point form assessment of the old and new cowboy myths concluded that the ranch hand of old needed to improve his image, his treatment of wildlife and women, and to change his clothing and his diet:

Old Myth:

- Cowboys are macho and simple-minded.
- Cowboys are 'men against nature'.
- Cowboys kill Indians.
- Cowboys leave women at home, wringing their hands.
- Cowboys are alone on the range.
- Cowboys are outlaws.

- Cowboys are brutal.
- Cowboys eat pork and beans.
- Cowboys dress for function.
- Cowboys shoot wolves.

New Myth:

- Cowboys are sensitive.
- Cowboys are advocates of the land.
- Cowboys are Indians.
- Women are cowboys.
- Cowboys ride the range with their wives.
- Cowboys have principles.
- Cowboys are kind.
- Cowboys eat grilled vegetable fajitas.
- Cowboys dress for functional style.
- Cowboys dance with wolves.

The assessment concluded that the latest incarnation of the American pioneer presented an opportunity for Marlboro to: 'Recapture the excitement that built the brand by recasting the Marlboro Man as the 90s Cowboy.'

Fortunately for the tobacco industry, the fight against anti-smoking propaganda was taken up by Hollywood. Although product placement had been found out and prohibited, the polarization that had occurred between smokers and non-smokers in real life had rendered the portrayal of smoking more or less obligatory in contemporary cinematic dramas. More than ever before, a character's tobacco habit – or opposition to the weed – could serve as visual shorthand and help to fix the character in the audience's minds. To the dismay of cancer surgeons everywhere, the traits Hollywood chose to attribute to smokers were independence, beauty and determination. A *USA Today* study in 1996 of eighteen current films, including the top ten at the box office, found only one, Disney's unforgettable *D3: The Mighty Ducks*, to be smoke-free. Another study of 250 of the most popular films of the decade showed that 87 per cent of movies contained tobacco use, nearly one-third showed tobacco brands as part of a scene or in a backdrop, and that a significant proportion of leading men and women, including Bruce Willis, Tom Cruise, Leonardo DiCaprio, Uma Thurman and Julia Roberts, smoked on screen. Furthermore, back in the 1980s when product placement had been legal, less than 1 per cent of cigarettes smoked could be identified by brand name, a figure which had increased to 11 per cent by 1997.

The revival of smoking in Hollywood movies was greeted with much hand-wringing among tobacco's opponents. Why were so many people smoking in films? Were they being paid in secret? A Hollywood adviser had a simpler explanation – edginess: 'smoking has become part of the definition of edginess. And edginess is in.' Hollywood also employed tobacco in its traditional role as an ambassador between the sexes. In the following extract from *Pulp Fiction* (1994), Vincent (John Travolta), a professional assassin, has been charged with looking after the wife of his gangster boss, the beautiful Mia (Uma Thurman) for an evening. We join the scene in Jackrabbit Slims, an imitation fifties diner.

MIA	Whaddya think?
VINCENT	It's like a wax museum with a pulse rate

Vincent takes out his pouch of tobacco and begins rolling himself a smoke. After a second of watching him –

MIA	What are you doing?
VINCENT	Rolling a smoke.
MIA	Here?
VINCENT	It's just tobacco.
MIA	Oh. Well in that case, will you roll me one, cowboy?

As he finishes licking it –

VINCENT	You can have this one, cowgirl.

He hands her the rolled smoke. She takes it, putting it
to her lips. Out of nowhere appears a Zippo lighter in
Vincent's hand. He lights it.

As Mia and Vincent smoke together the language between
them softens – the cigarette has once again worked its magic
between strangers.

Hollywood's example was followed in Great Britain, where
a resurgent film industry, fuelled by grants from the new
national lottery, also portrayed smoking in association with
glamour and heroism. UK film-makers tended towards realism
in their camerawork which obliged them to depict beautiful
women with cigarettes between their lips, because a majority
of this category smoked in real life. Whereas America led the
world in moving images of Divas, the UK and its European
partners were providing the latter day Helens – faces whose
pictures might launch a thousand ships – or at very least a
new perfume brand. Known collectively as 'supermodels',
these slender and enchanting women attracted endless attention from the media, and most of them smoked.

'You've come a long way, baby'

Lung cancer in old age is a hard argument to make to teenagers against fat in puberty. One of tobacco's oldest recorded virtues – its ability to suppress appetite – was enlisted in favour of the contemporary obsession with appearance. Young girls throughout Britain took up smoking in imitation of their idols, and to control their weight. Opponents of smoking attempted to counter this alarming development with temperance movement propaganda. The 'cigarette face' was revived. 'Smoking gives you wrinkles' said the posters. Smoking was also accused of causing bad breath and impotence. These arguments proved powerless against countless images of beautiful women with cigarettes between their fingers or their lips, who did not appear to have any wrinkles, and did not look as though they had bad breath. Further, such images were effective antidotes to pending impotence, and the anti-tobacco movement's blind wanderings into smoking's associations with glamour were clumsy failures.

The proportion of British women smoking rose in 1996, for the first time in more than twenty-five years. The increase was especially marked amongst teenage girls – by 1996 one in three 15-year-olds were smoking regularly. Not all these women took up smoking to resemble supermodels. The British women most likely to be smokers and least likely to give up were unemployed single women and lone mothers. Their habits were surveyed and published in a cheerfully titled HMSO publication: *When Life's a Drag – Women, Smoking and Disadvantage*. It appears that disadvantaged women did not smoke because they had been deluded by glamour, or inundated with advertising, but rather for the old-fashioned reason of relaxation. They smoked to calm themselves, and considered the act the only luxury in their lives. 'I smoke because everything I have ever wanted has been ruined. I haven't got anything else left now' (single mother).

The British government was aware of the plight of such smokers and even succeeded in summoning up a few pious observations on its most deprived citizens. Its 1999 White

Paper, 'Smoking Kills', confesses that 'in 1996, there were approximately 1 million lone parents on income support of whom 55% smoked an average of five packs of cigarettes a week at a cost of £2.50 a pack. That means lone parent families spent a staggering £357 million on cigarettes during that year', most of it in tax.

Despite the intensity with which it has been prosecuted, the anti-tobacco propaganda onslaught appears to have been fruitless, and has even been subjected to satire in the popular press. 'Michele lights up to celebrate No-Smoking Day' announced the *Daily Star* in 1998, and provided a picture of Michelle smoking in her underwear lest its readers were in any doubt. New arrivals in the British magazine market, spearheaded by *Loaded*, *FHM* and *GQ*, took to celebrating smoking culture, praising it in a manner not seen since the Victorian era. Smoking once again was portrayed as a rite of passage – a skill that every adult should understand, whether they smoked or not. Further, it appears that the public had become insensitive to health warnings, on the principle of 'cry wolf', as the following extract from American comedian Denis Leary's *No Cure for Cancer* illustrates. A smoker has just noticed that the size and severity of health warnings on his cigarette pack has been increased again: 'Yeah, Bill, I've got some cigarettes [*noticing warning on pack*] Hey! wait a minute. Jesus Christ. These things are bad for you! Shit, I thought they were good for you. I thought they had Vitamin C in 'em and stuff.'

The failure of the propaganda war against tobacco was lamented by the British government in 'Smoking Kills': 'survey evidence shows that half of all young people believe they have seen a cigarette advert on TV in the last six months despite the fact that it was banned 33 years ago'. The government concluded that the only way the battle can be won is through an outright advertising ban, for 'Any other measures we took would not work as well if they had to compete with stylish and powerful tobacco advertising.' Given that the 'stylish and powerful' advertising for one of the UK's leading brands (Silk Cut) consists of pictures of a piece of purple rayon with a slit in it, an outright ban may indeed be the government's only hope. This ambition received a millennial set-back when a proposed European Union-wide tobacco advertising ban (with the exception of Formula 1 sponsorship, after a timely donation to Great Britain's Labour Party won F1's management breathing space) was declared by the European Court of Justice to be unlawful. It appears that 'stylish and powerful' cigarette advertising showing neither cigarettes nor smokers may still be around to tempt unwary Britons in years to come.

Without doubt the principal beneficiaries of the controversy over smoking have been governments. Taking the British government as an example, the duties it received from the sale of tobacco amounted to more than £10.5 billion in 1997/8, accounting for roughly 80 per cent of the purchase price of a packet of cigarettes. From a financial point of view, therefore, the sale of cigarettes in the United Kingdom is principally a form of taxation and the main function of tobacco companies is tax collection.

A percentage of such taxation is justified as relating to the costs of treating dying smokers. In 1999, the most optimistic, in the sense of most inflated, estimate of such costs was £1.7 billion, leaving an excess income of £8.8 billion – enough to pay for all the abortions carried out in Britain during the decade, and the entire cost of Sizewell B nuclear power station. A further small deduction may be made towards the cost of government propaganda and programmes intended to dissuade tobacco users from their habits, though the sum is insignificant when set against the income stream. The government's White Paper 'Smoking Kills' announces, with some fanfare, that 'we are going to balance high tobacco tax with real support from the NHS to help smokers quit'. The 'real support' constitutes an investment of £60 million in a new NHS programme. How this pittance 'balances' £8.8 billion is not explained.

The British government also benefits financially from the tendency of smokers to die young. By getting it over with early instead of lingering on pensions or clogging up nursing homes in their wheelchairs, smokers provide their rulers with substantial savings. According to official figures, a smoker can expect to live sixteen years less than a non-smoker, which amounts to an average saving in pension costs, television licence discounts, bus passes and winter fuel allowances of c. £250,000 per smoker. It follows that it would be financial madness for the British government to ban smoking, and unless a better argument than its official estimated death toll of 120,000 smokers per annum can be found, smoking is unlikely to be prohibited in the British Isles.

The picture in the United States, which, historically, has not placed an excessive tax burden on smokers, is different. Only in the last quarter of the twentieth century have tobacco taxes at either federal or state level begun to approach the rates of other developed countries. The first significant rise took place in California in 1988, which created something resembling the gold rush of 1849 amongst anti-tobacco activists. Taxes were then increased piecemeal in other states. By 1994 they had reached an average of 52 cents per pack. The anti-smoking movement pressed for increases to $2 per pack, counting on support from the White House, whose new incumbent, President Clinton, had identified cigarettes as one product it may be permissible to overtax: 'I think health related taxes are different . . . I think cigarettes, for example, are different. We are spending a ton of money in private insurance and government tax payments to deal with the health care problems occasioned by bad health habits – and particularly smoking.' However, the most thorough survey into America's health-related costs of smoking, 'the cost of poor health habits', which had also taken into account the benefits of smokers dying young, estimated when converted to 1995 values that the net cost to American society of smoking was 33 cents per pack, i.e. less than existing taxes.

The legal posse of the anti-tobacco movement did not lose heart in the face of statistics – it reminded itself that there is more than one way to skin a cat, and that if the US government appeared reluctant to penalize tobacco companies further, the courts might not display the same restraint. The second group of winners to profit from mankind's affair with tobacco is comprised of American lawyers. Their eventual triumph in the US courts against the tobacco companies was also a victory of sorts for Dr Freud, whose theories were the first to cast addicts as victims of circumstance, and the largest legal damages awards in global history stem directly from the dreams enjoyed by a Viennese medical student after searching eels for their elusive testicles.

Tobacco companies started losing law suits in America in the 1990s. The first breach in their defences had occurred in 1988, when a jury awarded the family of Rose Cipollone, a dead smoker, $400,000 in damages. The award was later voided, but a principle had been established – a smoker could be a helpless victim, robbed of willpower as a consequence of their addiction, and if it could be shown that ciga-rette manufacturers knew their product was addictive, then they could be liable to compensate dying smokers and the relatives of dead ones. A number of US states, led by Minnesota and Florida, commenced exploratory proceed-ings against American cigarette manufacturers. They demanded access to the companies' internal documents and the companies complied with over 35 million pages. Perhaps they had resolved that if they were to confess, the action should be complete. In any event, the documents were a treasure trove of self-incrimination. As one of Minnesota's legal team observed: 'By the end of the day we couldn't believe what we were reading. It totally verified what we suspected.'

After confession came absolution. In 1998, the tobacco industry reached a Master Settlement Agreement with various American states which committed it to making payments in perpetuity to cover all the states' medical expenses incurred in treating sick smokers. The total cost to the industry over a twenty-five-year period will be $246 billion. According to the terms of the agreement, the industry will also set up and finance a foundation to reduce teenage smoking and investigate diseases associated with tobacco

use. It has also agreed not to target underaged smokers and
not to use cartoon characters in its advertising. All the individual companies involved in the settlement have also set
up websites, containing all trial documents and thereby
displaying their confessions to the world. The Master
Settlement Agreement will be paid for by raising the price
of cigarettes, i.e. by smokers.

While American smokers will benefit from the knowledge
that a significantly higher proportion of the price they pay
for cigarettes than UK smokers will be used for their benefit,
a sizeable proportion will benefit a very small number of
American anti-tobacco lawyers, presumably, all non-
smokers. By May 2000, over $10 billion in legal fees had
been awarded to the states' attorneys, with plenty more to
follow. In Mississippi, for instance, lawyers who represented
the state in the litigation surrounding the Master Settlement
Agreement with the tobacco companies will receive $1.4
billion between them, representing 35 per cent of the total
to be paid to the state for the care of sick and dying smokers.
In Florida, the fees work out at $233 million per lawyer,
causing Judge Harold J. Cohen to observe that the size of
such awards 'shocks the conscience of this court'. Even this,
however, pales into insignificance when placed beside the
fees that, pending appeal, may be awarded to Stanley
Rosenblatt, a Miami attorney, who recently won $145 billion
on behalf of sick Florida smokers. Rosenblatt took on the
case because he saw it as 'essentially a battle of good against
evil, of the exploited against the exploiters'. Rosenblatt's law
practice's share of the damages might make him 'richer than
Bill Gates'. Why start Microsoft when you can sue tobacco
companies?

The 1990s were a terrible decade for smokers in the United
States. Some took refuge in the new wonder drug Prozac,
which was considered less risky than cigarettes by the
medical profession. In addition to being encouraged to take
alternative mood altering substances, and to being circumscribed as to where they could enjoy their tobacco habits,
smokers were also condemned as lunatics by Harvard

medical researchers, who concluded that 'Americans with mental illness are nearly twice as likely to smoke cigarettes as people with no mental illness.' By extrapolating their results to the US population, the researchers estimated that people with diagnosable mental illness comprised nearly 45 per cent of the total tobacco market in the USA, thus implying that almost half of American smokers do so because they are insane. Perhaps the ultimate insult offered to an American smoker in the 1990s was the refusal of a last cigarette to death row inmate Larry White, before his execution in Huntsville, Texas, on the grounds that it would have been bad for his health.

There were a few rays of sunshine for the United States' oldest industry. Ex-President William Jefferson Clinton added a further use for tobacco to the existing list. As the Starr Report on which President Clinton's impeachment for perjury was based revealed: 'according to Ms Lewinski, "he focused on me pretty exclusively" kissing her bare breasts and fondling her genitals. A one point, the President inserted a cigar into Ms Lewinski's vagina, then put the cigar in his mouth and said "It tastes good".' *Per curiam,* a suspect has been identified for *that* cigar: the appropriately named Kinky Friedman, an American country and western star, whose hits include, 'They Don't Make Jews like Jesus Anymore', has recalled that Clinton 'invited me to the White House. I gave him a Cuban cigar – a Monte Cristo No. 2 [length, 6^{1}/$_{8}$th inches, ring gauge 42]. I told him: "Remember, Mr President, we're not supporting the economy, we're burning their fields." He laughed and took the cigar, but didn't smoke it – and that was before the Monica Lewinski story broke.'

Tobacco's position in other developed countries is more complex than in the UK and USA. The deadly menace of cigarette advertising, for instance, has not been countered with the same determination outside the Anglo-Saxon world. Tobacco advertising on Japanese television was finally and reluctantly banned in 1999, the government having too large a share of the cigarette market to consider it before. Both France and Spain, whose ancient monopolies recently

emerged to form 'Altadis', the sixth largest tobacco manufacturers in the world, have proved unreceptive to litigation
from dying smokers. The tragedy, in legal terms, of a state-
owned monopoly is that it is hard to differentiate from the
state, and as a consequence most French and Spanish
smokers currently dying have no other cigarette manufacturers than their governments to sue. Although smoking rates
have fallen in each country, market share is being won by
American brands, eager to pay those bills back home. As a
consequence, idiosyncratic French cigarettes, like idiosyncratic French cars, are disappearing from the French countryside. The Boyard, a black tobacco cigarette rolled in maize
paper, which had once ruled in the provinces, has become
rare as its aficionados have died, sometimes of old age. The
world's oldest person outside biblical times, Jeanne Calment,
coincidentally the world's oldest smoker, died in her bed in
Arles, in the south of France, in August 1997 at the age of
122 years, five months and two weeks.

Both France and Spain pushed up cigarette taxes during
the 1990s, resulting in a renaissance of smuggling in
Carmen's Andalusian homeland. Even though Spanish cigarette tax is low by European standards, cigarettes are no
worse for being cheaper, and the duty free ports of Gibraltar
and Ceuta clear astonishing numbers of cigarettes, as does
the little Pyrenean kingdom of Andorra, whose citizens are
officially the highest consumers of cigarettes in the world.
UK tobacco exports to Andorra rose from 13 million cigarettes in 1993 to 1,520 million in 1997, which works out
at average of 24,000 cigarettes per head per annum, and
is either a tribute to the lung power of the 63,000 men,
women and children of Andorra, or to their commercial
skills.

Perhaps one-third of all internationally traded cigarettes
are smuggled. Tobacco smuggling is a sizeable global
industry, whose revenues number tens of billions. In some
nations the activity is a testimony to the quality and consistency of the modern cigarette. Where domestic currencies
are weak, cigarettes still function as an alternative to cash,

the basic unit of exchange varying between a single cigarette in some African nations to a case of fifty cartons in
Thailand's Golden Triangle or the jungle strongholds of
Colombian cocaine dealers. The trade in smuggled cigarettes
is large enough for manufacturers to identify and compete
for market share. As Kenneth Clarke MP, non-executive
director of British American Tobacco, explained to the UK
press: 'Where any government is unwilling to act or their
efforts are unsuccessful, we act, completely within the law,
on the basis that our brands will be available alongside those
of our competitors in the smuggled as well as the legitimate
market.' Actions taken by tobacco companies have included
facilitating the availability of cigarettes to smugglers, sometimes by building warehouses close to borders with poor
customs control. Even government-owned tobacco companies, fearful of missing opportunities, have participated in
'General Trade' as the market for smuggled cigarettes is
known. For example, Japanese International Tobacco, a
state-owned behemoth with more than half of the Japanese
market, sells most of its cigarettes into Taiwan via the UK
and Switzerland using the General Trade route to circum-
vent an embargo on imports ex-Japan. Eighty-six per cent
of Japanese International Tobacco's sales volume in Taiwan
in 1993 was General Trade.

Tobacco is the contraband of choice all over the world.
Real camels carry loads of their namesake along the ancient
silk route. Cigarette adverts with the useful reminder
'Available in Duty Free' appear in countries where the
product is not officially available anywhere else. Tobacco
smuggling is a growth industry in the First as well as Third
World, and possibly even in the Second World, though this
category of nation receives less attention than the other two.
'The effectiveness of continued tax increases is being under-
mined by tobacco smuggling' laments the British government, bewailing the morals of its subjects who buy tobacco
on cross-Channel shopping expeditions and who 'may well
be prepared to sell tobacco to children'. Clearly these are
crocodile tears – no evidence has been adduced to prove

that untaxed cigarettes are greater killers than those on which a government has taken its cut. Meanwhile, British smugglers face jail sentences of up to seven years, during which they might perhaps learn to sharpen their trade, as cigarettes remain the unofficial currency of the UK's prisons.

The great British smuggling revival, reminiscent of the days of Good Queen Bess and strange King James I, had captured nearly 10 per cent of the kingdom's tobacco trade by the beginning of the third millennium. As a principal in the market, 'Trader Dave', explained to London's *Evening Standard*: 'I strongly promote bootlegging in the hope that it will eventually bring taxes down in England. I'm an ardent believer in equality and fairness. I've made a lot of money out of being the opposite, but I didn't cause this situation, and making all this money was an accident.'

Considering mankind's infatuation with the weed on a global basis, it appears that the ancient symbiosis is prospering as never before. People's knowledge of the health risks of smoking is partial at best, especially in low- and middle-income countries. In China, for example, 61 per cent of adult smokers surveyed in 1996 believed that cigarettes did them 'little or no harm'. A similar picture prevails in India, the world's third largest tobacco producer, where, according to a WHO report, 'Smoking is a status symbol among urban educated youths, but most appear to be unaware of the hazards of smoking.' Help, however, is at hand for the 65 per cent or so of Indian men who use tobacco.

Monkeys have once again risen to the occasion to fight evil. Revered in India for assisting Hindu god Ram in defeating demon king Ravan, monkeys are now fighting the evil of smoking with their very own set of rules. A group of 14 monkeys can be found marching in a single file around the Sahara India building on Boring Road in Patna, keeping a strict vigil on smokers. The moment they see anyone smoking they pounce on the person and deliver a couple of tight slaps. Only the other day a handful of monkeys surrounded a man who was smoking and took out all the cigarettes in his pocket. They then started smoking themselves and then coughed and whooped much to the surprise of the onlookers . . .

In South Africa, by way of contrast, some primates have taken up the tobacco habit, to the displeasure of the non-smoking owner of Natal Zoological Gardens, who explained his chimpanzees 'will smoke if someone throws them a cigarette, but I do not approve'. The chimpanzees were taught how to smoke, and to inhale, by one of their number who had worked previously in an American ice show. As the zoo's owner observed of the first example of other species adopting and disseminating a tobacco habit: 'I think these chimpanzees are proud of their ability to smoke . . . but I believe everything is OK if taken in moderation.'

It is doubtful that tobacco companies will consider other species as a market with much potential. Asia, especially China, represents tobacco's future. China is home to nearly one-third of the planet's 1.2 billion smokers, and its pattern of tobacco consumption is developing along familiar lines. Contemporary Chinese adolescents appear to have the same aspirations as their American counterparts of thirty years before: 'Zhang Fei, a 16-year-old senior high school student, has been smoking cigarettes for a year. On average, he goes through half a pack a day. Gazing at the smoke curling slowly from his mouth, he says: "Men look cool when they're smoking." . . . Likewise, Yang Tao, a stylish 16-year-old school drop-out extols the wonders of lighting up. "A lot of movie stars are smokers," she says. "Women look sexy when they smoke." ' In the face of such powerful associations, the director of the local tobacco control movement, the Committee for Patriotic Public Health, compares his efforts to the 'salt sprinkled over the sea'. Anti-tobacco programmes in China are further complicated by the folklore which has accumulated around tobacco over the centuries, entrenching reasons for smoking in Chinese culture. According to an old Chinese saying familiar to smokers and non-smokers alike: 'This life is better than the immortals' when you smoke tobacco after every meal.'

Tobacco is certanly the most equivocal substance in daily human use. According to the World Health Organization, 'The tobacco epidemic is a communicated disease. It is communicated through advertising, through the example of smokers and through the smoke to which non-smokers – especially children – are exposed. Our job is to immunize people against this epidemic.' However, to the 1.2 billion smokers of the world, tobacco is not just a killer, but a pleasure, a comforter and a friend. For over five centuries, tobacco has been integrated into cultures as diverse as mankind itself, each of which has evolved justifications for using the weed, some ancient, others original. Although tobacco has lost most of its religious associations, many of the oldest reasons for smoking are still in use, and still valid. Tobacco has recently been discovered to protect against some of the most devastating ailments of old age, including Alzheimer's disease and Parkinson's disease. It also has been shown to guard against cancer of the womb. As to tobacco's association with contemplation and thought, there are centuries of precedents of eminent smokers, who cannot be dismissed as simply victims of their ages or habits. Many great men and women have left elegant testimonies to their tobacco habits, which will be joined, I believe, with others made in centuries to come.

Appendix 1

How to Grow Tobacco

Modern Cuban tobacco producers liken their work to being married to a delectable but immoral woman who provides a single exquisite pleasure at the cost of endless hardship. Tobacco, however, is easy to grow on a non-commercial scale, and home-grown, with a modicum of care, will produce a satisfying smoke. The entire process, from planting seeds to final harvest, takes approximately twenty-four weeks. British subjects wishing to grow tobacco will need to organize their activities around the summer months. They will, however, find tobacco growing rewarding: a plantation the size of a double bed will provide enough tobacco to supply the average British smoker for a year at an equal cost (excluding sunshine and labour) to a packet of cigarettes. Those wishing to undertake production on a commercial scale are referred to the European Union, which dispenses handsome grants to tobacco growers in various member states, especially Italy and Greece. Tobacco, per acre, is the most highly subsidized crop in Europe.

This guide is intended for home production only. Commercial production and sale of tobacco is governed by strict legislation, and is subject to taxation in most countries. The primary law applicable in the UK is the Tobacco Products Duty Act 1979, as amended. HM Customs and Excise Notice 476 exempts home-grown tobacco from duty, provided that:

- you have grown the tobacco yourself;
- the tobacco products are for your own consumption; and
- you do not sell the products

Cultivation

The cultivation phase to topping and suckering takes six to eight weeks

The tobacco seedlings should be planted in rows at intervals of 50 cm, the rows to be spaced 1 metre apart. Each seedling should be placed in a hole, which is to be filled with soil so that the seedling's stem is covered with soil to a point 4 cm above the root. After transplanting, the seedlings should be lightly watered for two days in succession, then left for a week. They may appear wilted during this period – this however is 'transplant shock', a condition suffered by many plants subsequent to a sudden change in location, and will help the plants to become stronger.

For the first four to six weeks after transplanting the plants will grow slowly and begin to stand upright. They require only occasional watering, but as much sunshine as possible. They should be fertilized two weeks after transplantation, and again after six weeks, although if the plants are dark green and vigorous at the six-week stage they will not require more fertilizer. The tobacco bed should be kept free of weeds, and the soil built up into a little mound around the base of each plant.

After six weeks the plants should be about 50 cm high and about to commence their rapid growth stage. For the next fortnight or so the plants will require regular watering and observation. Despite tobacco's reputation as a killer, an enormous range of garden pests are very fond of it, and growers will need to protect their plants against infestations. This is most easily achieved by spraying the plants with insecticide weekly until topping and suckering. Tobacco plants are also susceptible to fungal infections, including moulds, stem rot, brown spot, frog eye and the ominous wilt. The plants may be safeguarded by spraying with fungicide fortnightly prior to topping and suckering. Finally, squirrels are known to become addicted to tobacco, some, such as those in Hyde Park in London, through eating discarded cigarette butts. Squirrels may be controlled by trapping or shooting.

Topping and suckering

Topping and suckering occurs seven to nine weeks after transplantation

Topping and suckering involves the amputation of the tobacco plants' flowers and suckers as they appear. Flowers and suckers would otherwise divert energy and nutrients from the leaves. The flowers will appear in the form of a bud at the top of the plants seven to nine weeks after transplantation. They should be pinched or cut from the plants immediately. Virginian tobacco farmers often grow their thumbnails long and sharp to assist them in this process. The suckers are new shoots which will appear at the point where the leaves join the stalk of the plant. They too must be pinched or cut away immediately.

The plants should be growing quickly at this stage. Depending on the strain, they should be 100 to 120 cm high ten weeks after transplantation.

Picking

Picking commences ten to eleven weeks after transplantation

Tobacco growers who have progressed successfully to this stage will have the opportunity to discover why tobacco is considered to be a labour intensive crop. The average Cuban tobacco plant is visited 170 times in its life, as its leaves must be picked singly, and over a period of time.

Each plant should have between fifteen and twenty tongue-shaped leaves. Those at the bottom of the plant will become 'ripe' first. This state can be determined by the appearance and feel of a leaf. Ripe leaves will have lost colour, turning from dark green to pale green or even yellow, and will have become less fleshy to the touch. Unripe leaves will lack flavour, so it is better to wait for the colour change before harvest. When possible, tobacco should be picked in dry weather.

While tobacco resembles food in that it stimulates the taste buds and satisfies hunger, unlike food, it is not cultivated as a source of nourishment, and tobacco growers should therefore resist harvesting too early, when leaves may appear ripe to eat, but will not yet have developed their full smoking flavour.

The first pick should harvest the four or five leaves at the bottom of the plant. The second pick, two weeks later, should take leaves from the middle of the plant. Subsequent picks, at approximately two week intervals, will take leaves from the middle and top of the plant, until it has been completely stripped. Picking should be completed sixteen weeks after transplantation.

Curing

In order for tobacco to have life after death it must be cured, which ensures that the organoleptic qualities of the weed are fixed. Harvest marks the start of the ceremony. The curing process, during which the tobacco leaves are fermented, then dried, is the most important single factor in the flavour and narcotic and toxic effects of tobacco smoke. Curing can last years and be composed of dozens of complex treatments. For example, the *Ligero* and *Seco* filler leaves used in the construction of a Cohiba cigar enjoy three separate fermentations, during the last of which they are packed in wooden barrels for eighteen months. The following is a guide to rudimentary curing, using only commonly available household articles, which will produce satisfactory tobacco for use in pipes and cigarettes, and as snuff.

After each pick, stack the leaves on top of one another, wrap in a double layer of plastic bags and place in the sunshine or by a heater to stimulate fermentation. Fermentation removes unsavoury nitrogen compounds, reduces tar and nicotine levels, and assists in creating uniform colour and texture. Fermentation should take four to five days, during which the leaves will 'sweat', i.e. exude moisture. When fermentation is complete, the leaves will have changed colour to yellow/brown, have shrunk in size and be silky to the touch.

Next, the fermented tobacco must be dried. Commercial growers usually carry out this process in barns heated by flues, and tobacco kilns for home drying which work on similar principles may be purchased via the Internet. However, air drying is a simple and effective method of completing the curing process, and should be carried out as follows: tie the tobacco leaves by their ribs to clothes-pegs. Weather permitting, hang them on a line in sunshine, otherwise, hang them in a warm location, sheltered from strong winds and rain. This will take about five days. Leaves should not be left out of doors at night, or in a place where they might become saturated with dew.

When the ribs of the leaves snap if bent double on themselves, yet retain a degree of elasticity, the tobacco is ready for consumption. All that is needed now is a keen eye and a sharp knife. Lay half a dozen tobacco leaves flat on top of one another, roll the pile from one edge to the other into a cigar shape, and slice thin cross-sections from the rolled leaves. Toss the tobacco shreds together and store in a pouch for use. Cigarette tobacco should be sliced as finely as possible, whereas that destined for combustion in pipes may be of a coarser cut. Growers wishing to make snuff should cure their tobacco slightly drier, then grate the rolled up leaves. The grated tobacco may then be dried over a heater, and pulverized in a mortar and pestle to produce a finer powder.

Tobacco not required for immediate use should be wrapped in a plastic bag and stored in an air-tight container in a dark place. Properly stored tobacco will last for two or three years.

Notes

Abbreviations

Calder – Angus Calder, *Revolutionary Empire*, Pimlico, 1998

Corti – Count Corti, *A History of Smoking*, trans. Paul England, Bracken Books, 1996

Dunhill – Alfred Dunhill, *The Pipe Book*, The Lyons Press, 1999

Ellis – John Ellis, *World War II – the Sharp End*, Windroe & Greene, 1990

Goodman – Jordan Goodman, *Tobacco in History*, Routledge, 1993

Hilton – Matthew Hilton, *Smoking in British Popular Culture 1800–2000*, Manchester University Press, 2000

Kluger – Richard Kluger, *Ashes to Ashes*, Vintage Books, 1997

Mackenzie – Compton Mackenzie, *Sublime Tobacco*, Chatto & Windus, 1957

Parkman – Francis Parkman, *The Oregon Trail*, Project Gutenberg Etext

Wilbert – Johannes Wilbert, *Tobacco and Shamanism in South America*, Yale University Press, 1987

1 To Breathe is to Inhale

General: examples of South American tobacco use derive from Wilbert, whose comprehensive examination of tobacco use and ritual in the region is recommended to those with a further interest in these areas.

6 *the sufferer's soul*, Goodman, p. 23.
6 *he who overcomes*, Wilbert, p. 156.
6 *filled with hungry*, Wilbert, p. 159.

2 Confrontation

21 *(it) does not*, S. E Morrison, *Journals and Other Documents on the Life and Voyages of Christopher Columbus*, New York, Heritage Press, 1963, p. 286.
23 *with a little*, Christopher Columbus journal entry for 6 November 1492 (Las Casas).
24 *These two Christians*, Bartolomé de Las Casas, *History of the Indies*.
25 *I and my companions*, Hernán Cortés, *Letters from Mexico*, trans. Anthony Pagden, Yale University Press, 1986, pp. 45–35.
25 *We will do*, Cortés, *Letters from Mexico*.
26 *Among other evil*, Gonzalo Fernandez de Oviedo y Valdes, *Historia general y natural de las Indias*, 1535.
26 *I have known*, Las Casas, *History of the Indies*.
27 *They venerate and*, Gonzalo Fernandez de Oviedo y Valdes, *Historia general*.
29 *there were also*, Bernal Dias, *The Conquest of New Spain*, trans. A. P. Maudsley, Da Capo Press, 1996, p. 211.
30 *the thirteenth day*, Cortés, *Letters from Mexico*, p. 265.
31 *There groweth also*, *The Voyages of Jacques Cartier* (1545), trans. John Florio, 1580.
32 *a finely wrought*, E. G. Bourne, 1907, quoted in Wilbert, p. 16.
32 *We found . . . the most*, Amerigo Vespucci, 1504.
33 *There is another*, André Thevet, *The New Found World*, 1568.
33 *It is hardly*, Girolamo Benzoni, *La Historia del Nuevo Mundo*, 1565.
35 *in perfect order*, William H. Prescott, *The Conquest of Peru*, Random House, 1998.
35 *We were amazed*, Cortés, *Letters from Mexico*.

3 Transmission

4 Bewitched

75 *whereas it is agreed*, quoted in Mackenzie, p. 113.

75 *Judgement is coming*, John Winthrop, *General Observations for the Plantation of New England*, 1628.

75 *There is no doubt*, John Davys, *The Seaman's Secrets*, 1594.

81 *I cannot refrain*, quoted in Corti, p. 98.

5 Middlemen and Foreign Bottoms

89 *they have some very*, Guy Tachard, *A Relation of the Voyage to Siam*, 1688, White Orchid Press, Bangkok, 1981, p. 72.

89 *The Hottentots, being*, Tachard, *Voyage to Siam*, p. 69.

90 *The poor sort*, quoted in Mackenzie, p. 226.

92 *modelled after the*, Dr John Fryer, 'A New Account of East-India and Persia', Hakluyt Society, *John Fryer's East India and Persia*, 1912, vol. II.

95 *it is both godless*, quoted in Corti, p. 115.

95 *We have graciously*, quoted in Corti, p. 158.

97 *This day, much*, Samuel Pepys, *Diary*, 7 June 1665.

97 *in England, when*, Jorevin de Rochefort (1672), trans. *Antiquarian Repertory*, vol. II.

98 *It is all spirit*, Robert Herrick, *The Tobacconist*.

98 *Here, all alone*, George Wither, 1661.

99 *down to Winchcombe*, Pepys, *Diary*, 19 September 1667.

101 *[the] Irish take*, James Howell, *Letters*, 1650.

102 *I find that*, quoted in Mackenzie, p. 210.

102 *who was sent*, quoted in Calder, p. 254.

104 *The Glasgow vessells*, Daniel Defoe.

107 *large quantities of*, J. C. Robert, *The Story of Tobacco in America*, Knopf, 1952, p. 99.

6 Magic Dust

110 *they import so,* quoted in Calder, p. 334, also Peter Kolchin, *American Slavery,* Penguin Books, 1995, p. 11.

112 *Callabars are not,* South Carolina planter Henry Laurens, 1757, quoted in Kolchin, p. 66.

113 *all servants imported,* Maryland Statute, 1692.

117 *The world has,* Johann Heinrich Cohausen, *Lust of the Longing Nose,* 1720.

120 *Is it only sold,* Giacomo Casanova, *History of My Life,* trans. Willard R. Trask, Johns Hopkins University Press, 1997, vol. III, ch. 8. Casanova's favourite snuff box, given to him by the lovely nun MM, had a false bottom which concealed a picture of her naked beneath a portrait of her in her nun's habit.

121 *with unparalleled resolution, et seq.,* quoted by Mackenzie, p. 167.

121 *I cannot see,* Richard Steele, *Tatler,* no. 101.

125 *How long has, Vintner's and Tobacconist's Advocate,* quoted in Mackenzie, p. 180.

126 *it is the highest,* Adam Smith, *The Wealth of Nations,* Penguin Classics, 1986, Book II, ch. 3.

7 Come the Revolution

128 *we proceeded for, et seq.,* Joseph Banks, *Endeavour Journal,* August 1768–1771, e-text from University of New South Wales.

132 *one day they,* quoted in Alan Morehead, *The Fatal Impact,* Hamish Hamilton, 1966, p. 43.

133 *I endeavoured to,* quoted in Morehead, *The Fatal Impact,* p. 78.

133 *One page of Hawkesworth,* 'An Epistle from Mr Banks, Voyager, Monster Hunter and Amoroso'.

134 *they may appear, The Journals of Captain Cook,* Penguin Classics, 1999.

134 *happy people, content,* Banks, *Journal.*

137 *These debts had*, quoted in Calder, p. 329.

137 *a rage for regulation*, Edmund Burke, *History of the Reign of George II*, quoted in Paul Johnson, *A History of the American People*, Weidenfeld & Nicholson, 1997, p. 134.

138 *The Americans must*, Earl of Chatham (Pitt), quoted in Johnson, p. 141.

139 *Men will speculate*, quoted in Johnson, p. 125.

141 *His clothing was*, the Comte de Segur, quoted in Johnson, p. 166.

143 *thirteen United States*, this exchange is related in Johnson, p. 170.

8 Bandoleros, Scum and Dandies

146 *In the count's room*, Leo Tolstoy, *War and Peace*, 1866. Tolstoy had a love-hate relationship with tobacco. Himself a heavy smoker, he dissuaded others from the habit, and eventually succeeded in renouncing it.

153 *By his bulk*, Charles Lamb, 'The Triumph of the Whale'.

153 *heroically consecrated to*, Thomas Carlyle.

155 *Sublime tobacco!*, George Gordon, Lord Byron, *The Island*, 1823, Canto II, verse xix.

156 *You don't mind*, William Thackeray, *Vanity Fair*, 1848.

157 *It did one's heart*, quoted in Arthur Bryant, *The Age of Elegance*, William Collins Sons & Co. Ltd, p. 224.

157 *I peeped into*, quoted in Bryant, *The Age of Elegance*, p. 234.

158 *not a good*, quoted in Mackenzie, p. 203.

159 *was considered the*, Mackenzie, p. 240.

159 *in a style*, H. J. Nellar, *Nicotiana* (1832), quoted in Mackenzie, p. 247.

160 *we must mention*, Brockhaus, *Konversationslexicon* (1809), quoted in Corti, p. 221.

160 *it is indescribable*, Police report, 23 July 1809, quoted in Corti, p. 225.

9 The American Preference

165 *The far reaching, Democratic Revue*, 1838.
165 *Land enough! Land enough!*, Major Daveznac (1844), quoted in Johnson, p. 379.
166 *a great Powhatan*, quoted in Johnson, p. 338.
167 *thorough savages, Neither*, Parkman.
167 *A large circle*, Parkman.
168 *we found the*, George Catlin, *Letters and Notes on the Manners, Customs and Conditions of the North American Indians*, Dover Publications, 1973.
170 *The lodge* of, Parkman.
171 *They smoke a*, Richard Henry Dana, *Two Years Before the Mast*, 1896, pp. 115–116.
174 *As Washington may*, Charles Dickens, *American Notes*, Everyman, 1997, p. 121.
175 *and if they happen*, Dickens, p. 130.
176 *there was as*, Mark Twain, *The Adventures of Huckleberry Finn* (1884), Penguin Classics, 1985, p. 201.
176 *I saw in*, Dickens, p. 143.
178 *She was wearing*, Prosper Merimée, *Carmen*, trans. Nicholas Jotcham, Oxford World Classics, 1998, p. 201.
179 *this immense harem*, Pierre Louÿs, *The Woman and the Puppet*, 1902.
180 *I am a writer's*, Charles Baudelaire, *Les Fleurs du mal* (1857), trans. Richard Howard, Everyman's Library, 1993, p. 107.
181 *In filial reliance*, Otto Gennerisch, *St Valentine's Day* (1843), quoted in Corti, p. 237.
182 *Franklin's fellow citizens*, quoted in Corti, p. 240.
183 *This cigar business*, quoted in Corti, pp. 244–5.

10 A Class is Born

185 *The tobacco used, Tobacco*, 10 (November 1890), quoted in Mackenzie, p. 266.
188 *He only smoked, et seq., The Life and Letters of Charles*

Darwin, edited by his son Francis Darwin, vol. I, Murray, 1887.

189 *swaddled carefully in*, Dunhill, p. 236.

191 *would be as rash, et seq.*, J. M. Barrie, *My Lady Nicotine*, Hodder & Stoughton, 1890, p. 27. Per curiam, the Arcadia Mix has been identified as Craven Mixture, a blend which still exists, which readers may be tempted to test for themselves to see if they are worthy.

191 *For Maggie has*, Rudyard Kipling, 'The Betrothed', 1888.

192 *Is it conceivable*, Wilkie Collins, *The Moonstone*, 1868. Collins was a fervent supporter of smoking: 'when I am ill . . . tobacco is the best friend that my irritable nerves possess. When I am well, but exhausted for the time by a hard day's work, tobacco nerves and composes me.'

193 *Off tumbles the*, Barrie, p. 36.

193 *that chamber of*, Ouida, *Under Two Flags* (1867), Oxford University Press, 1995, p. 18.

194 *When I could*, Mackenzie, p. 1.

195 *Not only is*, quoted in Hilton, p. 61.

195 *livid with impotent*, the papers of John Fraser, University of Liverpool Special Collections, 1105, quoted in Hilton, p. 46.

197 *A lone man's*, Charles Kingsley, *Westward Ho!*, 1855.

198 *As a rule*, Sir Arthur Conan Doyle, *The Red Headed League*, Book Club Associates edition, 1982, pp. 54–5.

199 *In former days*, quoted in Mackenzie, p. 310.

200 *It was my*, Henry M. Stanley, *Through the Dark Continent*, Dover Publications Inc., 1988, pp. 84–5.

201 *induced a highly*, R. C. K. Ensor, *England 1870–1914*, Oxford University Press, 1938.

202 *The Chancellor of*, quoted in Hilton, p. 30.

202 *wasted in the most*, Ensor, p. 20.

11 Automatic for the People

204 *miserable apology for, et seq.*, quoted in Hilton, p. 28.
208 *featured either photographs*, quoted in Goodman, p. 102.
215 *a beardless resemblance*, Kluger, p. 38.
216 *Many a young man*, J. H. Kellogg, *The Living Temple*, 1903.
217 *If our beloved*, J. C. Murray, *Smoking, when Injurious, when Innocuous, when Beneficial* (1871), e-text.
217 *generally turned into, et seq.*, Lieutenant General R. S. S. Baden-Powell, *Scouting for Boys*, 1909.

12 Of Cats and Camels

219 *Gentlemen, you may smoke*, no traceable source exists for these four words; they are however repeated by many tobacco historians, whom I will join.
221 *we would consider*, quoted in Kluger, p. 44.
222 *All it wants*, quoted in Kluger, p. 47.
225 *a violent action*, Henry Ford, *The Case Against the Little White Slaver*, Detroit, 1914, rev. edn 1916.
227 *even in my dreams*, quoted in Paul Ferris, *Dr Freud – A Life*, Pimlico, 1998, p. 35.
228 *Woe to you*, quoted in Ferris, *Dr Freud*, p. 35.
228 *who retained a*, Sigmund Freud, *Three Essays on the Theory of Sexuality*, 1905.
230 *Morphine is the, et seq.*, Ford, *Little White Slaver*.
232 *A large YMCA*, Siegfried Sassoon, *Memoirs of an Infantry Officer* (1930), Faber & Faber Ltd, 1965, p. 133.
233 *I could hear*, Sassoon, *Memoirs*, p. 21.
235 *On my left*, Edward Gibbons, *Floyd Gibbons – Your Headline Hunter*, Exposition Press, 1953.

13 Stars

238 *constant movement of*, Religious Tract Society (1898), quoted in Hilton, p. 143.
238 *they are simply*, L. Linton, 'The Wild Women (part II, conclusion): As Social Insurgents', *Nineteenth Century*, 30 (October 1891), quoted in Hilton, p. 143.
242 *a gentleman should*, Emily Post (1872–1960), *Etiquette*, 1922.
251 *a glance at*, Corti, p. 265.
252 *the most democratic*, *American Mercury* (1925), quoted in Kluger, p. 69.

14 Verboten

254 *wrath of the Red*, Robert N. Proctor, 'The Anti-tobacco Campaign of the Nazis – a Little Known Aspect of Public Health in Germany 1933–1945', Pennsylvania State University, 1996.
255 *extraordinary rise in*, interestingly, some of the Nazi research has been rehabilitated. Eberhard Schairer and Eric Schoniger's work was published for the first time in English in 2001 (*International Journal of Epidemiology*, 30, p. 24). The same issue carries a useful commentary by Robert N. Proctor.
255 *In London during*, quoted in Hilary Graham, *When Life's a Drag: Women, Smoking and Disadvantage*, Department of Health, HMSO, 1993, p. 7.
257 *I do not drink*, quoted in Anwer Bati, *The Complete Cigar Book*, HarperCollins, 2000, p. 223.
258 *It is permissible*, quoted in Anthony Beevor, *Stalingrad*.
258 *never smoked cigarettes*, quoted in Ellis, p. 295.
258 *had at one time*, quoted in Ellis, p. 295.
259 *a man who*, quoted in Ellis, p. 294.
259 *When he is*, quoted in Ellis, p. 48.
260 *Smoking boogers yer*, George MacDonald Frazer, *Quartered Safe Out Here*, Harvill, 1992, p. 146.

260 *Dig a hole,* Bill Maudlin, quoted in Ellis, p. 45.

261 *I myself had,* quoted in Anthony Beevor, *Stalingrad.*

262 *Of the four camps,* George Rosie, HQ. 3/506th, WWII, edited by Billy A. Carrington, CW3 (Retired), U.S. Army, thedropzone.org, 1999.

263 *We too grew tobacco,* quoted in Henner Hess, *The Other Prohibition – the Cigarette Crisis in Post-war Germany,* Johann Wolfgang Goethe Universität, Frankfurt.

263 *If we consider how,* Corti, p. 267.

266 *I offered him,* Douglas MacArthur, *Reminiscences,* 1964, p. 288.

15 The Spy Who Loved Me

270 *They still talk,* Nick Tosches, *Country: The Twisted Roots of Rock 'n' Roll,* p. 82.

274 *ought to make,* quoted in Kluger, p. 154.

277 *the most generally,* quoted in Kluger, p. 181.

278 *No one who lived,* Margaret Thatcher, *The Downing Street Years,* HarperCollins, 1993, p. 12.

278 *without them, life, Man and his Cigarette,* Mass Observation 3192, quoted in Hilton, p. 124.

279 *in public places, Man and his Cigarette,* quoted in Hilton, p. 125.

279 *a solitary indulgence, Man and his Cigarette,* quoted in Hilton, p. 125.

280 *Their actions, moreover, Man and his Cigarette,* quoted in Hilton, p. 130.

280 *The old Edwardian,* John Osborne, *Look Back in Anger,* Faber & Faber Ltd, 1957.

283 *I ask myself,* Mackenzie, p. 332.

283 *in the course,* Mackenzie, p. 343.

284 *This officer . . . remains,* Ian Fleming, *Thunderball,* Jonathan Cape, London, 1961.

16 Tumours

288 *research and public relations*, quoted in Kluger, p. 205.
288 *There are biologically*, Arthur D. Little, 15 March 1961, outside consultant to the Liggett & Meyers Tobacco Co. He conducted animal experiments from 1954–84. Bates No. 2021382496/2498 (also 2021382482/2509).
294 *the leadership in*, Sir Philip Rogers and Geoffery Todd, *Reports on Policy Aspects of the Smoking and Health Situations in the USA*, 1964.
301 *[W]e believe that*, 3 April 1970, Gallaher memo by company research manager to the head of Gallaher Ltd, American Tobacco's British-based sister company, Trial Exhibit 21,905, the memo is the subject of 'Amid Millions of Tobacco Documents, a Single One Stands Out', *Minneapolis–St Paul Star Tribune Electronic Telegraph*, 16 March 1998.
302 *Going out at night*, Michael Herr, *Dispatches*, Pan Books Ltd, 1978.
303 *The [soldier] lit*, Herr, *Dispatches*.
303 *Hunted, set traps*, Bao Ninh, *The Sorrow of War*, English version by Frank Palmos, translated by Vo Bang Thanh and Phan Thanh Ho with Katerina Pierce, Secker & Warburg, 1993.
304 *Tobacco was second*, Zhang Xianliang, *My Bodhi Tree*, Secker & Warburg, 1995.
304 *Of course the*, Xianliang, *My Bodhi Tree*.

17 Blowing Bubbles

308 *There is nothing*, Fidel Castro, *Letters from Gaol*, quoted in G. Cabrere Infante, *Holy Smoke*, Faber & Faber, 1985, pp. 73–4.
311 *It was on*, Peter Hillson to author, 1990.
311 *A cigarette is*, British American Tobacco, BATCO

researcher, Colin C. Greig, in a document thought to date from the early 1980s.

314 *Nobody has yet*, Dennis Potter in *The Sunday Times*, 30 October 1977.

314 *It is my conviction*, 13 February 1962, BAT memo, 'The Smoking and Health Problem', paper presented at Research Conference, Southampton, England, pp. 15–16, Trial Exhibit no. 11938.

315 *There is little*, Dr M. A. Hamilton Russell, 'The Smoking Habit and Its Classification', *The Practitioner* 212 (1974), p. 794.

315 *It requires no*, Russell, 'The Smoking Habit'.

316 *Younger adult smokers*, 29 February 1984, R. J. Reynolds document, 'Young Adult Smokers: Strategies and Opportunites'.

316 *McKennell (137) has examined, et seq.*, British American Tobacco, 'Dependence on Cigarette Smoking – A Review', Report No. RD153, 15 December 1977, Bates No. 105458896–105459086.

316 *True answers on*, Philip Morris, Bates No. 2048370187/0190.

318 *the level of nicotine*, Hamilton Russell, 'The Smoking Habit'.

321 *I've had the*, quoted in Kluger, p. 422.

324 *It was probable*, George Orwell, *1984*, 1949.

325 *cigarette smokers in*, from Andrew Whist, Vice President of Corporate Affairs, PM International, New York, to Bryan Simpson, Secretary General of INFOTAB in Brussels, 21 May 1984, Philip Morris document site: http://www.pmdocs.com/. Bates No. 2023272944/2945.

18 Man and Superman

328 *worth trying just*, quoted in Kluger, p. 460.

330 *The main thing*, quoted in Ralph Harris and Judith

Hatton, *Murder a Cigarette*, Gerald Duckworth, 1998, p. 24.

337 *Will have a*, quoted in Kluger, p. 711.

343 *a cigar held*, quoted in Corti, p. 257.

19 Hunde, Wolt Ihr Ewig Leben?

Title: Frederick the Great to his battle shy soldiers: 'Dogs, would you live forever?' quoted by John Updike, *Self-consciousness: Memoirs*, 1989, after giving up smoking for his health.

347 *The emergence of*, Philip Morris, 15 July 1992, Bates No. 2048360916/0937, from www.pmdocs.com.

349 *smoking has become*, Larry Deutchman, Entertainment Industries Council Inc., quoted in *USA Today*, 11 July 1996.

349 *MIA: Whaddya think?*, Quentin Tarantino, *Pulp Fiction*, Faber & Faber, 1994.

352 *in 1996, there*, 'Smoking Kills', White Paper on Tobacco, e-text, 1999, 1.24.

352 *Yeah, Bill, I've*, Denis Leary, *No Cure for Cancer*, Picador, 1992.

352 *survey evidence shows*, 'Smoking Kills', 6.4.

352 *Any other measures*, 'Smoking Kills', 2.5.

353 *we are going*, 'Smoking Kills', 2.18.

354 *I think health*, Clinton, quoted in Kluger, p. 735.

355 *By the end*, Roberta Walburn, Minnesota trial lawyer.

357 *Americans with mental*, Karen Lasser, Steffie Woolhandler, David Himmelstein, Danny McCormick, J. Wesley Boyd, David Bor, *Journal of the American Medical Association*, 22 November 2000.

357 *invited me to*, *Daily Telegraph*, 'Peterborough', October 2000.

359 *Where any government*, Kenneth Clarke MP, *Guardian*, 3 February 2000.

359 *The effectiveness of*, 'Smoking Kills', 6.50.

360 *I strongly promote*, *Evening Standard*, 7 August 1998.

362 *Smoking is a*, World Health Organization, *Tobacco or Health, a Global Status Report*, 1995.

362 *Monkeys have once*, India Express Newspapers (Bombay) Ltd, 2000.

362 *will smoke if, The Sunday Times*, 9 August 1998.

363 *Zhang Fei, a, South China Morning Post*, 16 October 2000.

363 *The tobacco epidemic*, Dr Gro Harlem Brundtland, International Policy Conference on Children and Tobacco, 18 March 1999.

Further Reading

BOOKS

Barrie, J. M., *My Lady Nicotine*, Hodder & Stoughton, 1890

Bati, Anwer, *The Complete Cigar Book*, HarperCollins, 2000

Cabrera Infante, G., *Holy Smoke*, Faber & Faber, 1985

Corti, Count, *A History of Smoking*, trans. Paul England, Bracken Books, 1996

de Sahagun, Bernardino, *Historia general de las cosas de Nueva España*

Dowling, Scott (ed.), *The Psychology and Treatment of Addictive Behaviour*, Monograph 8, American Psychoanalytical Association

Dunhill, Alfred, *The Pipe Book*, The Lyons Press, 1999

Goodman, Jordan, *Tobacco in History*, Routledge, 1993

Gordillo, Manuel Rodríguez, *Un archivo para la historia del tabaco*

Hakluyt, Richard, *Principal Navigations* (1600), Penguin Classics

Harris, Ralph and Hatton, Judith, *Murder a Cigarette*, Gerald Duckworth, 1998

Hilton, Matthew, *Smoking in British Popular Culture, 1800–2000*, Manchester University Press, 2000

Inventario general de autos de la renta del tabaco: Reino de Sevilla.

Kluger, Richard, *Ashes to Ashes*, Vintage Books, 1997

Lewis, Meriwether and Clark, William, *The History of the Lewis and Clark Expedition*, Dover Publications

Mackenzie, Compton, *Sublime Tobacco*, Chatto & Windus, 1957

Marin, Francisco Rodríguez, *La verdadera biografía de Nicolás Monardes*, vol. II, Editorial Padilla Libros , Seville, 1988

Marrero, Eumelio Espino, *Cuban Cigar Tobacco*, TFH Publications, 1996

McFarlane, Anthony, *The British in The Americas 1480–1815*, Longman, 1994

Monardes, Nicolás, *Historia medicinal de las cosas que se traen de nuestras Indias occidentales que sirven en medicina*, vol. I, facsimile edition, Editorial Padilla Libros, Seville, 1988

Ortiz de Lanzagorta, J. L., *Las cigarreras de Sevilla*

Parry, J. H., Sherlock, Philip and Maingot, Anthony, *A Short History of the West Indies*, Macmillan Caribbean, 1987

Perez Vidal, Jose, *España en la historia del tobaco*, Edt Consejo de investigaciones científicas, Centro de Estudios de Etnologia peninsular, Madrid, 1959

Porter, Roy (ed.), *Medicine, a History of Healing*, The Ivy Press, 1997

Relación acerca de las antigüedades de los Indias. Primer tratado escrito en América. (Romano Pane), Traducing by Jose Juan Arrom, Edt. Siglo XXI, 1974

Rudgley, Richard, *The Alchemy of Culture: Intoxicants in Society*, British Museum Press, 1993

Schele, Linda and Miller, Mary Ellen, *The Blood of Kings: Dynasty and Ritual in Mayan Art*, Thames & Hudson, 1992

The Journals of Captain Cook, Penguin Classics, 1999

Wilbert, Johannes, *Tobacco and Shamanism in South America*, Yale University Press, 1987

Reports

Advisory Committee to the Surgeon General of the Public Health Service, *Smoking and Health*, United States Department of Health, Education and Welfare, Washington, 1964

Chinese Snuff Bottles from the Collection of Mary and George Bloch, October, 1987, catalogue by Robert W. L. Kleiner

Department of Health, *Report of the Scientific Committee on Tobacco and Health*, HMSO, 1998

El Galeon de Manila, Exhibition Catalogue, Hospital de los Venerables, Seville, 2000

Master Settlement Agreement dated November 23, 1998

Royal College of Physicians, *Smoking or Health*, Pitman Medical, 1977

Société General, 'Tobacco Sector Analysis', June 2000

Index

Picture Credits

The Publishers have used their best endeavours to contact all copyright holders. They will be glad to hear from anyone who recognises their photographs.